T0160391

Witch's Garden

DISCLAIMER

Witch's Garden is intended for general informational purposes only and should not be relied upon as recommending or promoting any specific practice or method of treatment. It is not intended to diagnose, treat or prevent any illness or condition and is not a substitute for advice from a healthcare professional. You should consult your medical practitioner before engaging in any of the information detailed in this book. You should not use the information in this book as a substitute for medication or other treatment prescribed by your medical practitioner.

The publisher, Royal Botanic Gardens, Kew and the author make no representations or warranties with respect to the accuracy, completeness or currency of the contents of this work, and specifically disclaim, without limitation, any implied warranties of merchantability or fitness for a particular purpose and any injury, illness, damage, liability or loss incurred, directly or indirectly, from the use or application of any of the contents of this book. Furthermore, the publisher, Royal Botanic Gardens, Kew and the author are not affiliated with and do not sponsor or endorse any uses of or beliefs about plants as being scientifically accurate or advisable in any way referred to in this book.

Published in 2020 by Welbeck,
An imprint of the Welbeck Non-Fiction Limited,
part of Welbeck Publishing Group.
Based in London and Sydney.
www.welbeckpublishing.com

 A CIP catalogue for this book is available from the British Library.

ISBN 978-1-78739-436-0

Printed in Dubai

10 9 8 7

Witch's Garden

Plants in folklore, magic and traditional medicine

Sandra Lawrence

Introduction

What is a "witch's garden"? A garden owned by a person of magic – or a plot growing plants to ward off evil? For that matter, why do we associate the word "witch" with "evil"? For many people today, being a witch is a good thing, beneficial and joyful. In other parts of the world, witchcraft is still something to be feared.

However precarious life is today, for our ancestors it was even more dangerous, even more mysterious. Who knew what made humans healthy, successful, wealthy or loved? Everyone, from scholars to local practitioners, looked to the natural world around them for answers, and plants were obvious candidates for making sense of life. "Wise women", herbalists, apothecaries and witches worked in not dissimilar ways to early scholars, philosophers and scientists: empirically, testing out plants to see what they did, especially as medicines. The "cunning folk" also added a mystical element: explaining the role of plants in both spiritual and everyday life. Over thousands of years, complex folkloric traditions evolved, sometimes independently similar in different parts of the world.

The line between science and magic has always been thin; add religion into the mix and opinions – and emotions – begin to heat up. Plants became battlegrounds: as cultures waxed and waned, a herb "belonging" to a goddess might be reassigned to a saint – or to the devil. Boundaries became blurred. Astrology, for example, remained a serious point of consideration in medicine well into the Christian period. As the wealth of folklore grew, so too did the romance around it. We are not immune from that glamour today.

Folklore and superstition, both local and general, do not die easily, but they do change. Sometimes, properties of a plant can take on two diametrically opposed qualities according to region, culture and even individuals.

A book this size cannot begin to fully explore the complex and confusing history of herbs and the place they hold in our lives, even today. All it can do is dip a toe into the basic concepts of what makes plant lore so very fascinating. *Witch's Garden* looks at the ancient beginnings of practices and beliefs, some herbal pioneers and one or two of the "big ideas". It focuses on some of the most important plants, their historical uses and often contradictory associations, and is illustrated with beautiful pages from books and herbals kept in the archives at the Royal Botanic Gardens, Kew. Here's hoping it will inspire further reading and a deeper dive into the extraordinary world of plant lore.

Sandra Lawrence

Chapter 1: Plants of the Ancient World

Some of our most potent images of ancient civilizations involve plants: the lotus, beloved of the ancient Egyptians and found in wall paintings, architecture and even preserved whole in tombs; hemlock, the deadly poison drunk by Socrates, who was condemned to death for "corrupting" Athenian youth; olive branches, awarded as prizes at the world's first Olympic Games; and laurel wreaths, which crowned the emperors of Rome.

Respect for the earth's resources was essential to ancient life. Animals, minerals and plants were gifts from the gods, to be revered – and used – in every possible way.

Whether it was resin from the myrrh tree (*Commiphora myrrha*), used in Egyptian mummification; chocolate (*Theobroma cacao*), drunk for virility and power by the Aztecs; or aloe (*Aloe vera*) used by Native Hawaiians to soothe burns, uses of folk medicine such as these created a vast foundation of knowledge that grew over generations, and developed into a community database of cures.

Traditional Chinese Medicine (TCM) began well over 2,000 years ago. It combined many disciplines, including acupuncture, moxibustion (a form of heat therapy, practised by burning herbs near the skin), diet and herbal medicine and aimed to restore *qi*, a delicate balance of two opposing factors– *yin* and *yang*. According to TCM, every person has a slightly different composition and herbs are chosen carefully to maintain an individual's *qi*. Herbal remedies were classified into four natures, ranging from hot (extreme *yang*) to cold (extreme *yin*), and five flavours (acrid, sweet, sour, bitter and salty), each of which affects the yin-yang balance. Acrid herbs, for instance, were used to generate sweat, while bitter herbs purged the bowels. A third classification, meridians, controlled which part of the body the herbs worked on.

Kampo, while based on Chinese concepts, is uniquely Japanese. It has been practised for around 1,500 years and is still used alongside conventional medical practices.

In Hindu mythology, Ayurveda, the ancient medical system of South East Asia, was created by Dhanvantari, physician to the gods. Its earliest ideas came from the Vedas, a collection of religious texts dating from between 1500 and 1000 BCE. Like Chinese medicine, Ayurveda works on a system of balancing energy to maintain health, largely through diet, and many Ayurvedic preparations are plant-based. Sushruta, a surgeon from sometime around the sixth century BCE, took this one stage further. He used plants in procedures as well as in cures. Sushruta famously invented cosmetic or reconstructive surgery with the first nose job. His rhinoplasty involved taking a piece of skin from the cheek, measuring it carefully with a leaf, keeping the nostrils open with stems of the castor oil plant (*Ricinus communis*) and sprinkling the newly constructed nose with a powder made from liquorice (*Glycyrrhiza glabra*), red sandalwood (*Pterocarpus santalinus*) and barberry (*Berberis vulgaris*). The graft was then covered with cotton and lubricated with sesame oil (*Sesamum indicum*).

The ancient Greeks also embraced the idea of a healthy mind within a healthy body and their doctrine of the four humours strived toward both physical and mental well-being. Since they always

Opposite Sacred lotus (*Nelumbo nucifera*) after a painting by Peter Henderson for *The Temple of Flora* by Robert John Thornton, 1799–1807.

The Sacred Egyptian Bean

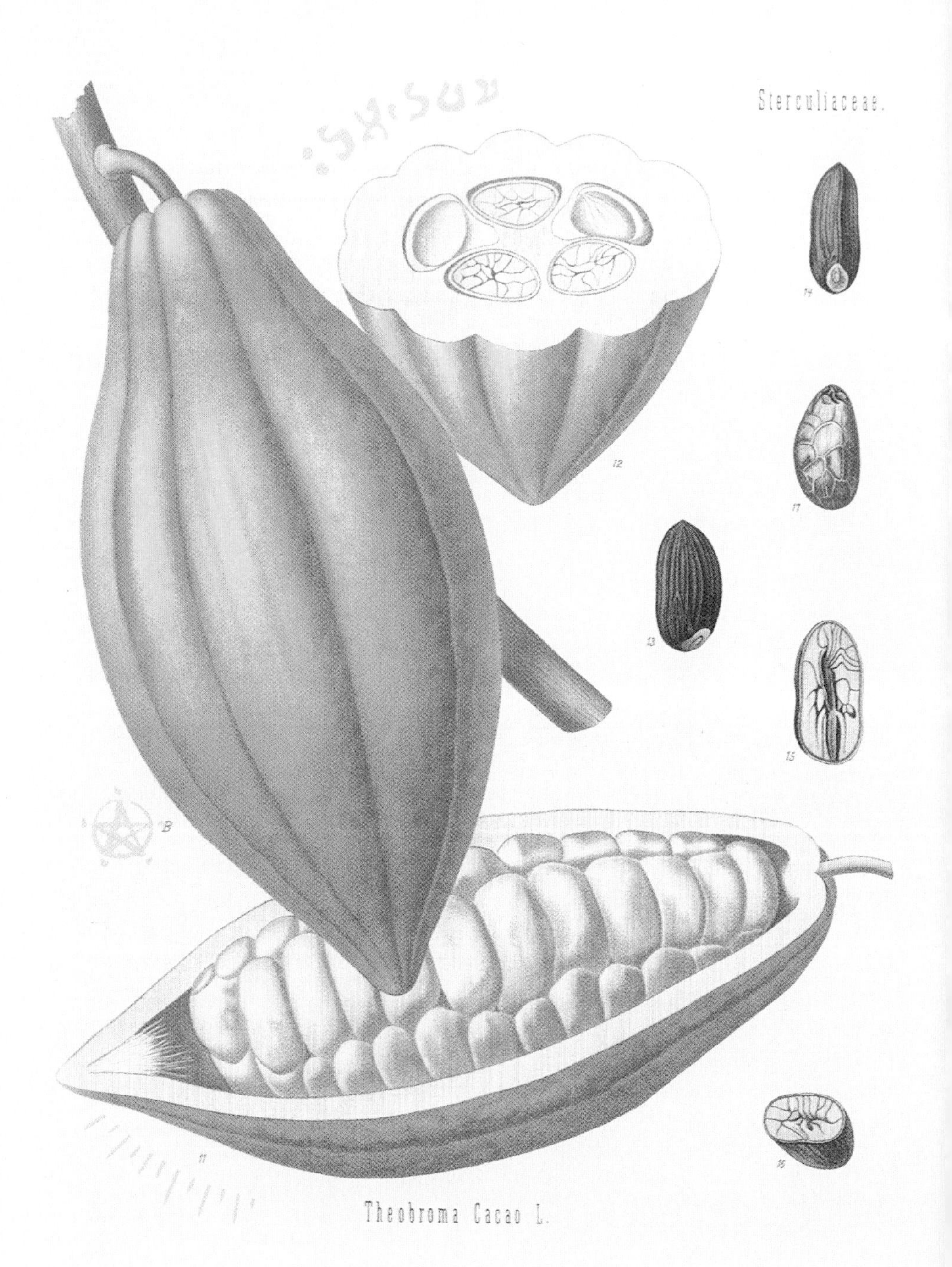
Sterculiaceae.
Theobroma Cacao L.

seemed to be at war with someone, medicine often revolved around battle wounds. The Olympic Games saw new uses of plants and their derivatives, such as applying olive oil to an athlete's skin before competing, to warm his body and so avoid injury.

Aromatic herbs were collected by specialist *rhizotomoi* "root-cutters", a secretive, closed-shop profession steeped in legend (the mythical witch Circe and centaur Chiron were both rhizotomoi). To protect their sources, root-cutters spread legends of the dire consequences of casually gathering herbs. Woodpeckers, they whispered, gouged out the eyes of anyone cutting peonies, while eagles protected hellebores and a mandrake's scream caused instant madness.

Like so many things, from architecture to philosophy, Roman medicine was heavily influenced by that of the Greeks – not that they always agreed with it. Pliny the Elder was unimpressed with Greek doctors, whom he felt overcharged and engaged in immoral conduct. He believed local, traditional Roman medicine, kept within each family, was better. Herbs played a major role. Pills (*pastilli*) often contained exotic ingredients such as saffron (*Crocus sativus*), pepper (*Piper nigrum*) and cinnamon (*Cinnamomum verum*), but more modest plants were also prized. Cato the Elder was a fan of cabbage (*Brassica oleracea*). Among dozens of other efficacies, he claimed it could draw foul pus from an ulcer, remove polyps from nostrils and even cure cancer. Babies bathed in the urine of someone who had eaten cabbage would never, he declared, be weaklings.

Opposite Cocoa (*Theobroma cacao*) from Köhler's *Medizinal Pflanzen*, 1887. Cocoa was drunk by the Aztecs before momentous events.

The Roman Empire was vast and explorers brought new herbal treatments from every land they touched. They also took their favourite remedies with them, which then became naturalized in their new environment. Alas, the widespread belief that stinging nettles were brought to Britain by soldiers, who flogged themselves with the hairy, acidic stems to keep warm in the freezing northern climate, is untrue. The plant is native to the UK. Romans were crazy for herbs in both food and medicine. Some herbs were considered so beneficial that they were literally eaten to extinction. We have no idea what an apparently extremely effective natural contraceptive, *Silphium*, even looked like, because the Romans gathered it to death.

For the ancient Celts, plants played a key role in food, daily life and medicine. It was in the interests of every Celt to know the general properties of plants, even if only to avoid poisoning themselves. If they cut themselves, figwort (*Scrophularia*) would allegedly staunch the blood, and they believed a case of intestinal worms could be cured with wild garlic (*Allium ursinum*). Today's scientists may explain the Celts' common-cold remedy – elderberry (*Sambucus*) – as good old-fashioned vitamin C.

Ancient Scandinavians, from the Vikings onward, used the specific properties of plants in left-field ways. In the twelfth-century tome *The Legendary Saga of St Olaf*, the hero Þormóður receives an arrow to the chest. A healing woman offers him a pungent broth of garlic and onion, figuring that if she can smell the soup through the wound, the bolt will have penetrated his vital organs and he will die. Þormóður refuses the broth, pulls out the arrow and dies anyway, but the idea was sound.

A number of now-illegal narcotics were first discovered – and used, either as religious

hallucinogenics or medicine – by ancient civilizations. Coca leaf (*Erythroxylum coca*) was chewed and made into a stimulant tea by the Maya. Harmal (*Peganum harmala*) was used as an antidepressant and to treat inflammation and fever in Iranian cultures, while images of magic mushrooms (*Psilocybin*) appear on frescoes in North Africa dating from between 9000 and 7000 BCE. Archaeologists recently discovered cannabis (*Cannabis sativa*) in the 2,500-year-old grave of a woman who, it appears, was using it to gain relief from breast cancer. The plant we know as the opium poppy (*Papaver somniferum*) is discussed by Homer in *The Odyssey*. Alas, their antiquity does not make these drugs any less dangerous.

Virtually all modern herbal medicine relies on the trial and error of our ancestors, who had the courage – or perhaps desperation – to observe and experiment, systematically putting plants in their mouths, wounds or other orifices in the hope of finding relief. History doesn't generally record the failures: individuals who discovered, the hard way, that nightshade (*Atropa bella-donna*) is deadly or that poison ivy (*Toxicodendron radicans*) can give as good as it gets. We owe them.

Above *Mount Fuji with Cherry Trees in Bloom*, woodcut print by Katsushika Hokusai, 1804. Thousands flock to Japan every year to attend the annual Hanami (flower-viewing) festival.

Opposite Herbarium sheet of myrrh (*Commiphora myrrha*), collected in Somaliland, 1932.

Sheet I
ABYSSINIA—SOMALILAND BOUNDARY COMMISSION.
Commiphora cuspidata Chiov.
(fide Chiov.)
Loc. Somaliland: Duwi, 2800 ft.
Coll.: J. B. GILLETT.
No. 4385
Date 20-10-1932

Hemlock

Conium maculatum

One of the ultimate "witches' plants", treacherous hemlock is traditionally associated with necromancy.

Shakespeare's witches toss "Root of hemlock, digg'd i' th' dark" into their cauldron and it is referred to earlier in *MacBeth* as the "insane root". The plant was said to be hallucinogenic, used in flying ointment and profoundly poisonous charms. Commonly found in hedgerows, and therefore easy for tweedy English murderers to source, hemlock insinuated itself into the twentieth century as one of Agatha Christie's poisons of choice. Victims suffer paralysis, loss of speech and death from asphyxia. Chillingly, the mind remains clear right up to the moment of death.

Vickery's Folk Flora lists more than 35 nicknames for poison hemlock, all of which conjure extraordinary images: "bad man's oatmeal", "scabby hands" "devil's blossom" and "break your mother's heart" are just a few. "Honiton lace" and "lady's needlework" refer to hemlock's curds of airy, lace-like flowers and feathery pinnate leaves. These are similar to those of cow parsley and are from the same Apiaceae family, which also includes carrots, celery and fennel. Unlike any of these, however, this plant is definitely *not* edible.

Hemlock is native to large areas of Europe and North Africa, but has naturalized across other continents, from Australia to North America. Like many members of Apiaceae, it is biennial, establishing in damp wastelands and ditches in its first year and then flowering and fruiting the following spring. All parts of the plant are poisonous. It is at its most potent when it is fresh, but even dried it is not to be trifled with.

Because of this, it was traditionally used only on external parts of the body, and then only in moderation – although even this is not to be advised now. Noting that the plant is ruled by the planet Saturn, Culpeper suggests roasting the root and applying to the hands to soothe gout and inflammation. One of the most enduring "cures" associated with hemlock was as an eye salve. Oddly, it was to be applied to the unafflicted eye.

Hemlock's most famous connotation lies with someone allegedly killed by it. There's no doubt the philosopher Socrates could have been put to death by this ancient Greek execution method (also said to have done for the out-of-favour politicians Theramenes and Phocion). Modern toxicologist Enid Bloch has studied Plato's account of Socrates's death and concluded that it does accurately describe peripheral neuropathy brought on by the alkaloids in *Conium maculatum.*

Opposite Herbarium sheet of hemlock (*Conium maculatum*), collected in Folkestone, UK, 1895.

UMBELLIFERÆ

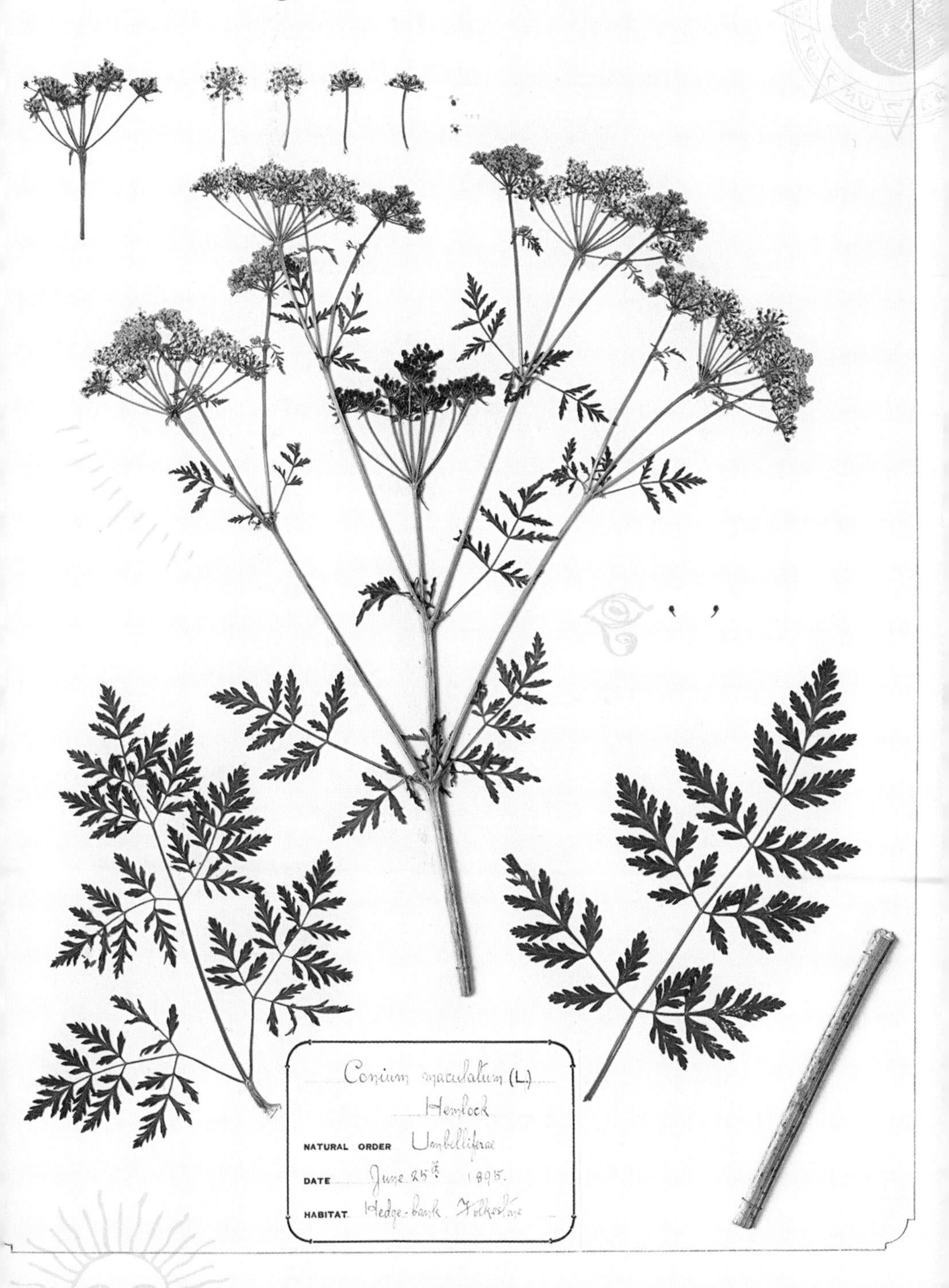

Black hellebore

Helleborus niger

Like many herbs later embraced by Christianity, black hellebore has a charming legend. A simple country girl wanted to visit the Christ child but had no gift to give.

An angel, touched by her piety, struck the ground and brought forth a flower. It's a good job the baby Jesus didn't put the plant, also known as "Christmas rose", in his mouth.

The ancients knew all about the dangers of black hellebore. When digging one up, it was important to ensure there were no eagles around. If one spotted you, death was sure to follow. Of course, this was just one of the dire warnings professional root-cutters put around to stop amateurs gathering their own herbs, but it isn't such a crazy notion. Black hellebore is one of nature's nasties. A powerful emetic, it is potentially fatal if ingested. Perhaps the first human to nibble a hellebore root was unfortunate enough to glimpse a bird of prey before keeling over.

If the correct rituals were performed, however, it was very useful. Pliny recommended drawing a circle around the plant, facing east and making a prayer before digging it up. Animals would be protected from evil spirits – and flies – if garlanded with hellebore. One superstition even whispered that the plant could render invisibility if scattered in the air in the correct manner.

Native to southern and central Europe and part of the Ranunculaceae or buttercup family, black hellebore grows in shady mountainous areas. We love it today for its drooping, five-petal flowers with purplish edges and yellow centres, but herbalists prized the roots. Its main medicinal use was as a purgative, used to induce vomiting after poisoning or eating bad food. Children were sometimes given hellebore to rid them of ringworm. Predictably, it didn't always end well.

Practitioners did at least acknowledge it was to be treated with respect. Hippocrates suggested patients rest and eat before taking the herb, then keep moving after ingesting it. Under no circumstances should they be allowed to sleep. He warned that any convulsions would be fatal. Other side effects, including diarrhoea and cardiac problems, as well as burning with prolonged skin contact, makes it surprising anyone thought it was a good idea, but hellebore was still being taken in eighteenth-century Britain. Botanist and herbalist Nicholas Culpeper recommended the herb for, among other things, leprosy, jaundice, sciatica and as a pessary to "provoke the terms [menstruation] exceedingly". Even this eye-watering use pales into insignificance against his recommendation for powdered hellebore: "strewed upon foul ulcers, it consumes the dead flesh".

Opposite Black hellebore (*Helleborus niger*).

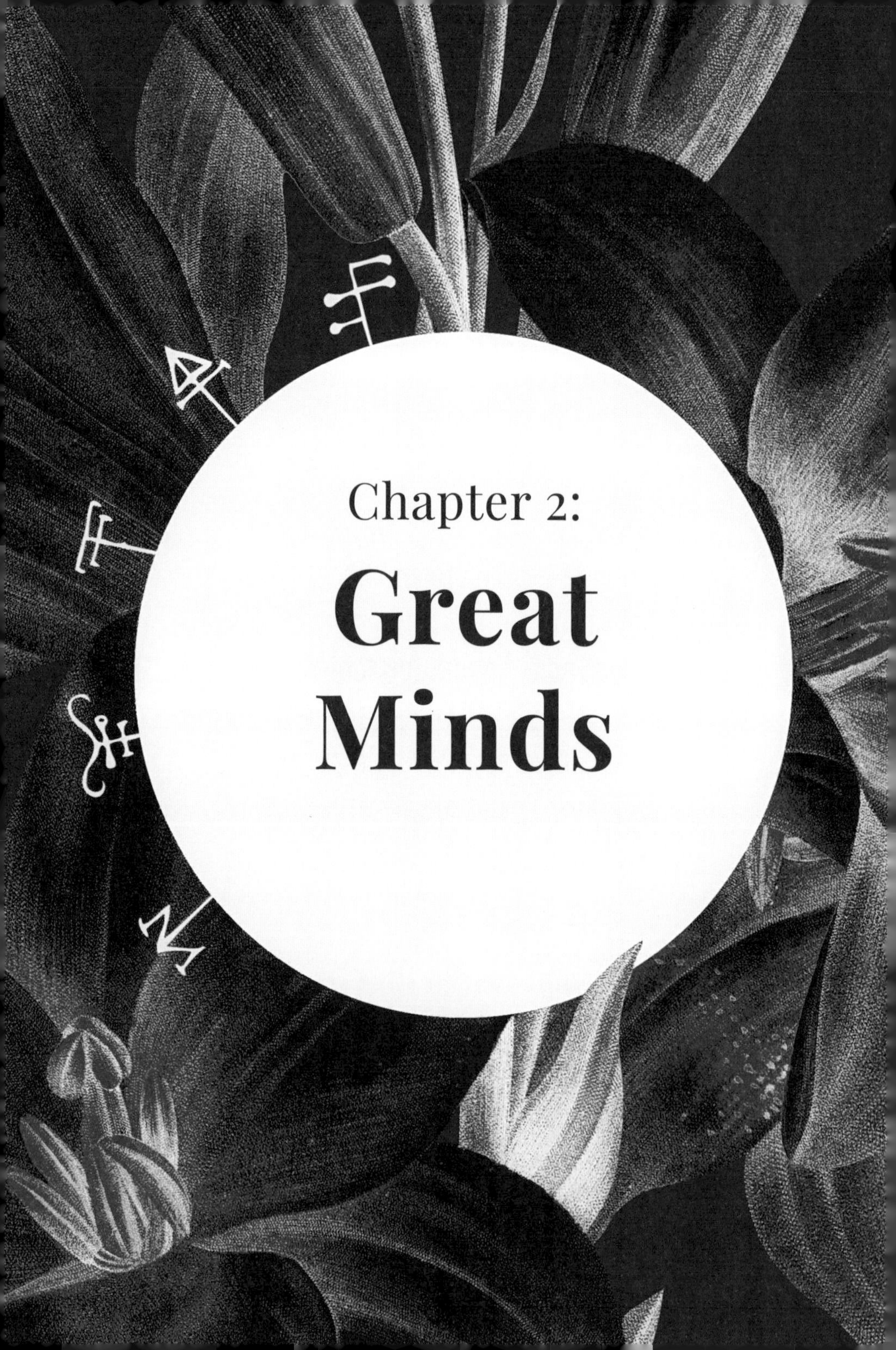

Chapter 2:

Great Minds

The history of herbal medicine is not easy to define. The first folk uses of plants, by humans from the times of mammoth-hunt injuries and flint-arrow wounds, have been lost to time. Written accounts are by no means the whole story, but they – and smatterings of archaeological evidence – are all we have.

Pilgrims still travel from around the world to Lieshan Mountain in China, to venerate Emperor Yan Shennong, father of Chinese agriculture, herbalism and medicine. "The Divine Farmer", as he was called, goes so far back in history that Shennong has acquired mythical status.

Second of three legendary emperors, he had the head of a bull and the body of a man. Among Shennong's many inventions were the cart, the plough, tea, the calendar and the practice of clearing land with fire. He was said to have a transparent stomach, and used his own body as a lab: he would ingest herbs, one by one, and observe what each did to him. As is the fate of so many self-experimenters, Shennong finally ate a flower that poisoned him and he died before he could find an antidote.

Whether or not Shennong existed, his catalogue of 365 species of medicinal plants certainly does. The *Shennong Bencao Jing* ("The Divine Husbandman's Herb Root Classic") is a manual of medicinal herbs, compiled in the late Han dynasty sometime around 306 BCE to 220 CE. It is still a must-read for students of traditional Chinese medicine.

In this text, herbs were graded on their rarity as well as their individual qualities representing the three realms: heaven, earth and man. Toxicity did not need to be avoided at all costs, just weighed against potentially beneficial outcomes. Poisonous herbs were to be combined with precise dosages of other drugs to counteract the bad aspects.

One of the first great minds of the West was Theophrastus of Eresus (approximately 372–287 BCE). A peripatetic philosopher and true Aristotelian, Theophrastus studied widely in subjects as diverse as physics, zoology, ethics, botany and the history of culture. Perhaps one of the reasons we now call him the "father of botany" is that two of his few works that have survived are *Peri phytõn historia* ("Enquiry into Plants") and *Peri phytõn aitiõn* ("On the Causes of Plants").

Theophrastus describes both wild and cultivated plants, their properties and practical uses, from his own research – the result of many years burning sandal leather – and reports from other travellers, including followers of Alexander the Great. His work would be hugely influential in the work of philosophers to come.

Gaius Plinius Secundus, better known today as Pliny the Elder (23–79 CE) wasn't really a "philosopher", he was far too hands-on in the world. A wealthy Roman military commander, he was naturally curious and used his various posts and political positions to study and write.

He is credited with seven books, but only one, *Naturalis Historia* ("Natural History"), remains complete. Its 37 volumes are a combination of older material gathered into a kind of encyclopaedia with observations made on his travels. Any gaps, Pliny filled in with legendary beasts or folklore, giving magic, superstition and science equal emphasis.

Naturalis Historia relies heavily on Theophrastus, and even if Pliny's translations are sometimes a little loose, the vocabulary is rich and much of

Opposite Herbarium sheet of *Camellia sinensis* – tea – collected in Ningbo, China, 1844.

HERBARIUM KEWENSE.
No:53.
CHINA : Ningpo
Camellia.
Coll. and presented by MR. W. HANCOCK. 1844.

what we know about Roman farming and gardening comes from him. Archaeological investigations are beginning to show that in this aspect, at least, his work is accurate.

Pliny's work is important in many ways. His Latin synonyms of Greek plant names made earlier Greek writings, including those of Theophrastus, identifiable. His many opinions on "modern life" – for example, his grumblings about the price of drugs and sharp practices by so-called physicians are both entertaining and illuminating, bringing ancient Rome to life and highlighting touching details others would have considered insignificant.

Alas, Pliny's curiosity would be his downfall. In 79 CE, as commander of the fleet stationed at the Bay of Naples, he went ashore to investigate some peculiar cloud formations around Mount Vesuvius. The famous eruption that buried Pompeii and Herculaneum did for the philosopher, too.

Many medieval monastic libraries contained a copy of Pliny's works. No one even thought to challenge him until 1492 when an Italian physician, Niccolò Leoniceno, pointed out some errors. However, Pliny remained important, known to Shakespeare and Milton, and his vibrant account of ancient Rome still shines.

We know little about the life of Pedanius Dioscorides (approximately 40–90 CE), one of ancient Greece's most important botanists. Born in Anazarbus, in what is now Turkey, he travelled as a surgeon with the Roman army. Wherever he went, he studied the properties of local plants and scoured local folk knowledge for new cures. His most famous work, *De Materia Medica* ("On Medical Substances"), was originally written in Greek but was soon translated into several languages, supplemented with commentary from European, Indian and Middle Eastern pharmacists. Written between 50 and 70 CE, it was not the first herbal compendium by any means, but vastly superior. In five books, Dioscorides describes nearly 600 plants and nearly 1,000 simple drugs. Some are chemical, such as mercury, lead and copper oxide (none of which feature heavily in modern medicine), but far more are concoctions, poultices and other dressings made from plants. Herbs are collated into therapeutic groupings: warming, binding, softening, drying, cooling and relaxing. Dioscorides lists plant names, synonyms, habitat, uses, properties and even the side effects of plants, as well as directions on how to gather, prepare and store them, along with warnings about how their purity may be adulterated if they land in the wrong hands.

For the next 15 centuries, Dioscorides's work was set in stone, but he was also an advocate of experimentation. Ultimately, his advice would be heeded and new research begun.

The monasteries of Europe kept copies – or, if not, many local adaptations – of the great Classical works, referring to them for their own therapies. St Hildegard of Bingen (1098–1179), a Renaissance woman before her time, would have been well versed in the writings of Theophrastus and Dioscorides as well as later writers, such as the ninth-century monk Walafrid Strabo of Reichenau, who wrote the early gardening book *Hortulus*. She just added her own touches.

Educated by Benedictine nuns, Hildegard became one herself at age fifteen. A poet and composer, Hildegard also made several books and is

Opposite Illumination from St Hildegarde von Bingen's theological work *Scivias*, *c.*1151. She influenced art, music, philosophy, theology – and the use of herbs in medicine.

credited with two treatises on medicine and natural history, *Physica* ("The Book of Simple Medicine") and the *Causae et Curae* ("Causes and Cures"). Although there is some argument by modern scholars as to how much she actually wrote of *Causae et Curae*, the inspiration behind it is certainly hers.

Hildegard advocated the concept of *viriditas*, or "greenness". She saw the natural world as an expression and celebration of God, rather than the traditionally evil vision of a fallen Eden soured by Satan. Even more revolutionary, she believed the "divine" was female in spirit.

She practised what we now call holistic healing, taking the ancient Greek concept of the four humours and adapting it so that patients would "rebalance" using a variety of therapies including hot baths, sleep, healthy diets, fasting, virtuous conduct and prayer. A herbal section of *Causae et Curae* discusses 500 plants, trees, stones, metals and creatures with healing properties. Hildegard travelled widely throughout Germany on the medieval equivalent of the literary speaking tour, but was only formally canonized as a saint in 2012.

John Gerard (approximately 1545–1612) was definitely not a saint. He was, however, a good gardener and a brilliant self-promoter. Born in Nantwich, England, in 1545, the Elizabethan barber surgeon became superintendent of Lord Burghley's garden and then curator of the physic garden at the College of Physicians.

Stirpium Historiae Pemptades Sex was a Latin herbal published in 1583 by Flemish botanist Rembert Dodoens. Robert Priest, of the London College of Physicians, began translating it into English, but he died before completing it and the completion of the job fell to John Gerard. He fiddled about with it, added some entries and presented the work as his own, complete with 1,800 woodcuts from an entirely different book (though he added 16 new ones, including the first-ever depiction of a potato). It contained a lot of errors, which botanist Matthias de l'Obel was invited to correct. Highly offended that his work might be anything other than perfect, Gerard dismissed l'Obel's (accurate) corrections and the book was published, mistakes and all.

Gerard's 1597 *Herball*, for all its dubious authorship, is hugely important to plant historians. Even more useful is a corrected version from 1633, in which the sloppy translations have been tidied up by Thomas Johnson, who diplomatically implies that Gerard meant well, but bit off more than he could chew.

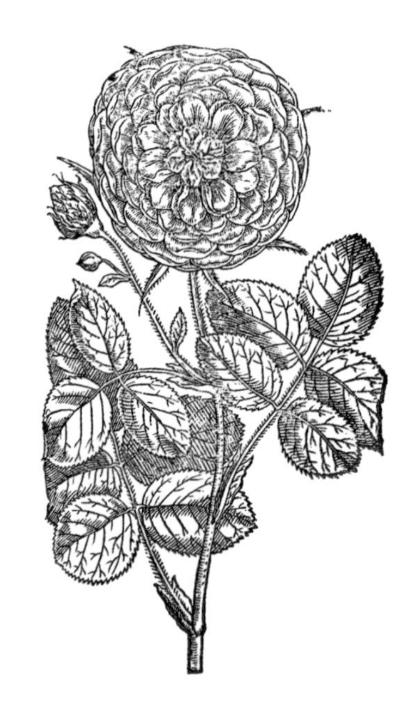

Above Woodcut of a rose from Gerard's *Herball*, 1597.

Opposite Frontispiece from Gerard's *Herball*, which scholars consulted well into the nineteenth century. It is still an important source for plant historians.

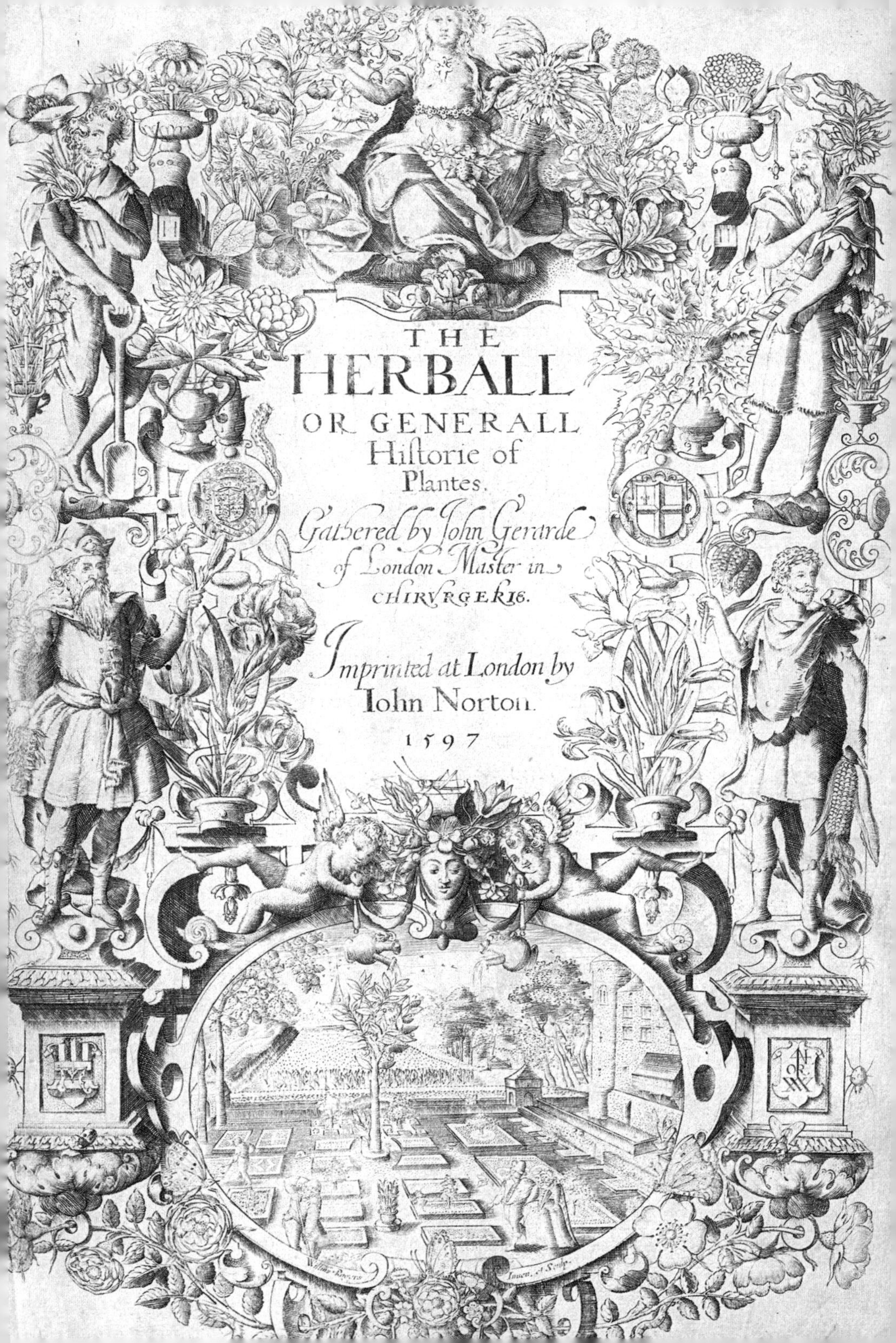
THE
HERBALL
OR GENERALL
Historie of
Plantes.
Gathered by John Gerarde
of London Master in
CHIRURGERIE.
Imprinted at London by
Iohn Norton.
1597

Blackberry

Rubus fruticosus

The humble blackberry is one of those rare plants still widely harvested by regular folk on afternoon rambles through the countryside.

Blackberrying, however, has not always been universally acceptable, even in times of relative hunger. In some areas, the plant's wood was thought to be that used in Christ's crown of thorns, and therefore the fruit was of the devil. Blackberry brambles were even planted on the graves of the dead to prevent them from walking.

Belief that the berries should not be eaten late in the year because the devil spits (or urinates on) them is, however unlikely, sound advice, as the fruit becomes prey to a fungus in the autumn. The date on which the devil spits moves chronologically, from the end of August to well after Michaelmas, and geographically, as ripening times move farther north.

Blackberry has always been an important hedging plant. Humans and small creatures appreciate the security value of prickly stems and the small creatures don't care what the devil's done to the berries – they taste good. The long, bending stems are semi-deciduous and will take root wherever they touch the ground, creating "arches". These were considered magical, powerful tools against a number of complaints. Horses "touched" by shrews, children with rickets or whooping cough and people plagued with blackheads needed to crawl through the hoop a few times, east to west, the way of the sun, and their troubles would be gone. Seven or nine times was generally considered a good amount. On the Welsh border, people left a slice of bread and butter as an offering. Whichever creature ate the bread would also take on the disease. Crawling through the magical hoop could also bring luck at cards, though would-be card sharps needed to understand that they would be entering into a pact with the devil.

Medicinally, the plant was popular. Both the Greeks and Romans treated gout with blackberry – and if ancient Greek physician Nicander of Colophon (approximately 197–139 BCE) recommending the flower as a cure for sea-monster stings sounds fanciful, Nicholas Culpeper, writing around 1,700 years later, was still suggesting it for serpent bites. He also liked it for the treatment of ulcers, putrid sores, bloody flux and "spitting of blood". The roots allegedly broke up kidney stones, while the leaves could apparently make a lotion for the mouth or "secret parts". Blackberry vinegar was a country medicine staple, a good cure-all for sore throats, coughs and chest ailments.

Opposite Blackberry (*Rubus fruticosus*) by Mary Anne Stebbing, 1946.

Blackberry
Rubus fruticosus

Zingiber officinale. Rosc.

Ginger

Zingiber officinale

Ginger is one of the superstars of herbal medicine, and has been riding high for millennia.

Native to India, China and South East Asia, the plant has leafy stems, bright green leaves and yellow-green flowers that can grow up to a metre tall, though it is most prized for its fleshy rhizomes, enjoyed in both food and medicine since the time of Confucius.

It is one of the great spices, traded by the very earliest travellers and a staple of Silk Road trade. The ancient Greeks baked it into their bread and sweetened gingerbread spread across the medieval fairs of Europe. It filtered through diplomacy (Queen Elizabeth I held a banquet where all the guests were depicted as "gingerbread men") and folklore (Hansel and Gretel's witch lives in a house of gingerbread). Young girls were encouraged to make gingerbread gifts to improve their chances of snaring a husband. Cookies included hearts, flowers and depictions of the young men in question. It's unlikely their chaperones would have been quite so keen if they'd known that the time-honoured Indian *Kama Sutra* text recommends ginger to arouse sexual energy.

The root's power as an aphrodisiac had long been noted, including by Dioscorides and Pliny, and it had a reputation for inflaming hearts in the Middle Ages. Given the nineteenth century's more prudish times, it's hardly surprising it later settled into slightly safer territory, leading to "general romance" rather than sex.

Perhaps ginger's hot reputation was, literally, its heat. Valued as "fire-giving", it was said to warm the body, soothing arthritis and stomach pains. It helped to sweat out fevers and colds. It could be administered raw, but it was fiercer dried. John Gerard noted that while candied ginger was hot and moist, dried it "heateth and drieth in the third degree".

Ginger was anti-inflammatory, relieved gassy stomachs and cleansed the body. Ginger tea was drunk to combat nausea, to improve circulation and, perhaps counterintuitively, to soothe burns. Nineteenth-century English innkeepers left powdered ginger on the bar for travellers to sprinkle into their beer so they could warm up after a hard journey. Less charmingly, unscrupulous horse traders "gingered up" flagging nags by applying the raw root to the creatures' anuses, making them "act lively".

In some traditions, ginger was a powerful ingredient in folk magic. It featured in spells to bring money, love and success – and eating ginger before working magic made the magician even more potent.

Opposite Ginger (*Zingiber officinale*) from Kohl's *Die officinellen Pflanzen der Pharmacopoea Germanica*, 1891–95.

Mint

Mentha

There are many different varieties of mint, used widely in cooking across the world since time immemorial.

It has even been found in Egyptian tombs. A small, sweet perennial herb, it has small flowers, pale pink or white. The aroma, however, is in the leaves, which can be anything from bright yellow-green to dark, almost black-green. Some varieties have hairy leaves, others smooth. Flavours range from mild, such as the garden spearmint (*Mentha spicata*) to the more fiery peppermint (*Mentha × piperita*).

An ancient Greek legend tells of Minthe, a river nymph who had caught the eye of Hades, god of the Underworld. Hades's wife Persephone took issue with this and turned Minthe into a plant that people would tread on. Hades couldn't undo the magic, but gave Minthe perfume so he could at least smell her.

Mint is, of course, still valued for its fresh smell, used in baths, perfumes and cool drinks. In ancient Athens, people scented distinct parts of their bodies with different fragrances; mint was used on the arms. It is naturally antibacterial, but that's not why the ancient Greeks used the herb to clean banqueting tables; the smell was considered a message of welcome. Many people would have welcomed the fourteenth-century use of mint as a breath freshener; it is still by far the most popular flavour of toothpaste. As John Gerard notes, "the smelle rejoyceth the heart of man".

Mint was also one of the classic medieval "strewing" herbs, laid on floors to discourage ants and disguise stale smells, but even then the idea was not new. The ancient Hebrews had scattered mint on synagogue floors, while the Romans used mint to stop milk from curdling – this anticoagulant theme continues down the ages. *The Good Housewife's Handmaid* of 1588 said the plant would clarify curdled milk, while Culpeper recommended it to disperse curdled milk in the breasts. He also noted it should never be given to a wounded man, as the lesion would not heal. Mint was also used for stomach problems, to soothe insomnia, anxiety, dizziness and flatulence.

Folklore dictates that mint should never be paid for. Ideally, it should be stolen from a neighbour. Happily, it grows easily and sends out runners. It is said to grow best in homes where the wife is dominant. Maybe that is why, in Mexico, it is known as *yerba buena* – the good herb.

Opposite Peppermint (*Mentha × piperita*) from Köhler's *Medizinal Pflanzen*, 1897.

Mentha piperita L.

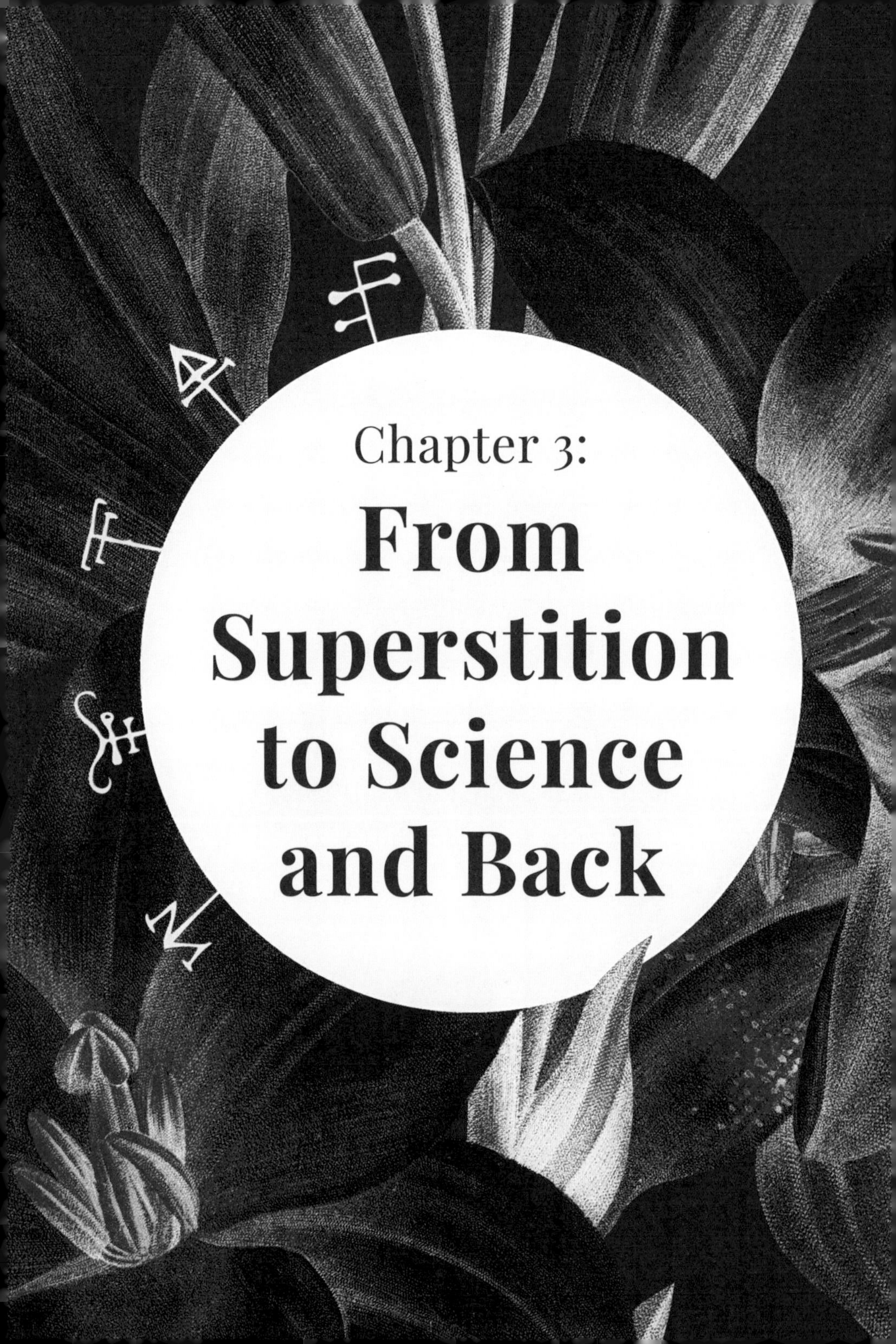

Chapter 3:

From Superstition to Science and Back

Health has been a large part of the job specifications for witch doctors, shamans, priests and faith healers since ancient times. A thorough knowledge of the plants available was essential to ensure successful work with the spirits.

Without scientific evidence as to why a herb sometimes might not work, people created elaborate rituals to go with physical medicines. These would, of course, only be guaranteed efficacious if practised precisely, by the right people.

There were so many unexplained phenomena around them that early scientists had no reason to doubt the existence of what we might call the supernatural. Sea monsters, lightning gods or ghosts were all just plain natural to them. Magic was as plausible as anything science could suggest and even the most empirical of scholars remained open-minded. Healing was inextricably linked with faith. In many ways, it still is – the placebo effect works, even occasionally, with people who know the pill they have taken is solely sugar.

By the middle of the first millennium, organized religions had grasped the herbal baton. Buddhist and Hindu institutions sought medical answers and Muslim scholars placed great emphasis on herbal medicine.

In the West, Christian monasteries were good places to continue research because the monks there could read Latin and Greek. Villages, towns and cities grew around individual monasteries or abbeys, as communities to service the everyday needs of the monks or nuns in residence. In return, the nuns and monks were expected to heal the sick people brought to their gates.

Some scholars argued that illness was a punishment from God. Others believed God had given them the provisions to treat illness and it was their duty to discover those cures. Monks studied classical works such as those of Dioscorides, Pliny the Elder and the Greek physician and surgeon Aelius "Galen" Galenus, copying the texts and adding their own discoveries. The tenth-century *Bald's Leechbook*, a collection of recipes from many sources, including Anglo-Saxon and ancient Greek and Latin writers, shows a remarkable knowledge of herbs and their uses.

The Rule of St Benedict, written sometime around 529 CE, insists special care must be taken of the sick to honour God. Many monasteries had gardens, but the Benedictines paid particular attention to horticulture, believing gardening tools were as important as chalices. Both monks and nuns were involved, including the seventh-century St Gertrude of Nivelles, patron saint of herbalists, gardeners and house cats, and St Fiacre, the Irish saint of gardening. The Benedictines perfected the art of making tinctures, which are herbal essences captured in alcohol.

Emperor Charlemagne so admired the Benedictine "physic" gardens that he ordered that all monasteries should include them. The Plan of St Gall is an architectural drawing, made around 820 CE, of a Swiss Benedictine monastery. It includes an infirmary, a bloodletting house and a garden, complete with suggested herbs.

Not everyone could get to a monastery. Wise women (occasionally men) were the bearers of medical knowledge within the secular community.

Opposite Coloured ink drawing of a pomegranate resembling the human jaw, *c.*1923, after G.B. Della Porta. An example of the doctrine of signatures.

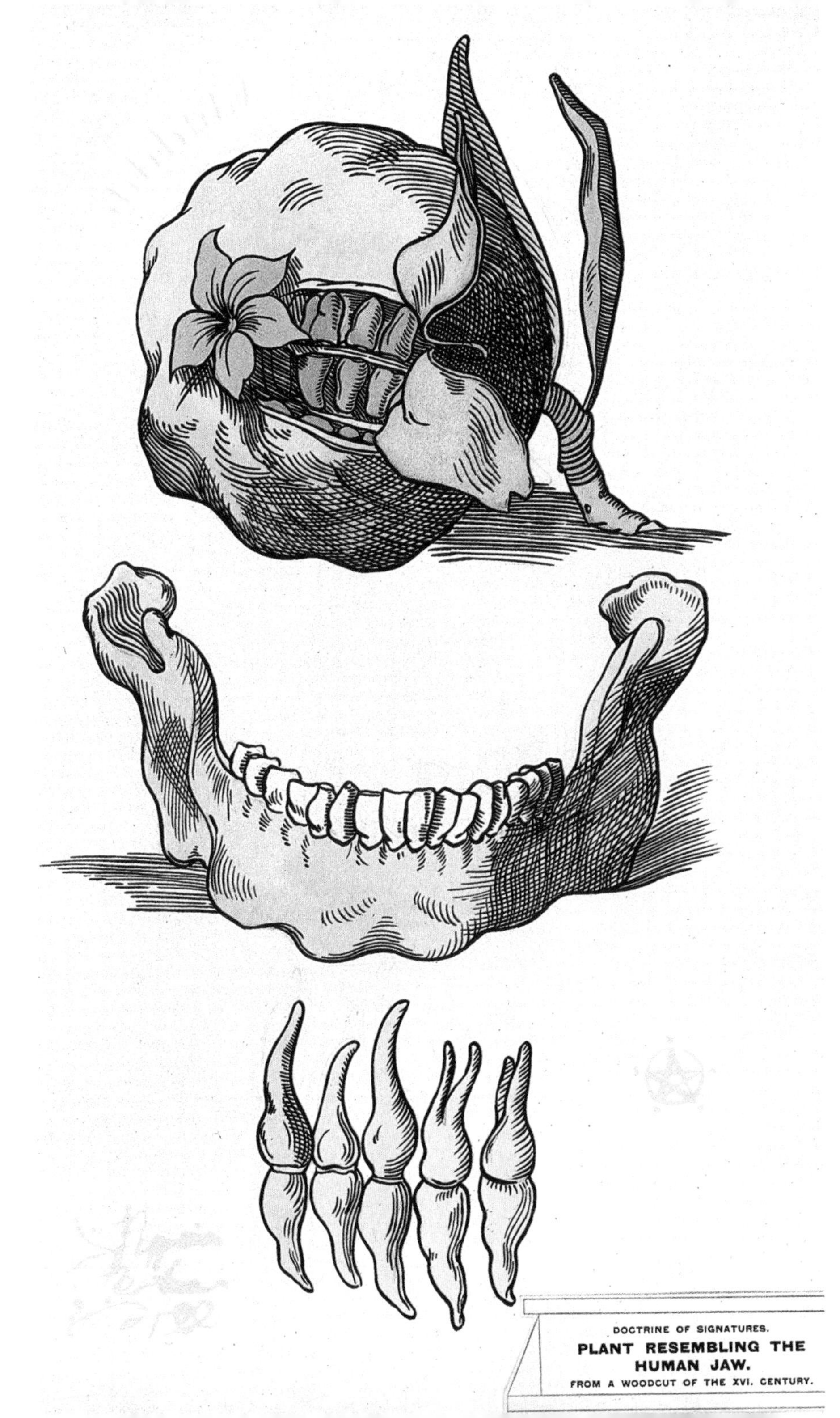

DOCTRINE OF SIGNATURES.

PLANT RESEMBLING THE HUMAN JAW.

FROM A WOODCUT OF THE XVI. CENTURY.

Above Seventeenth-century print of a Roman "plague doctor", who wore beaked masks stuffed with what were considered protective herbs.

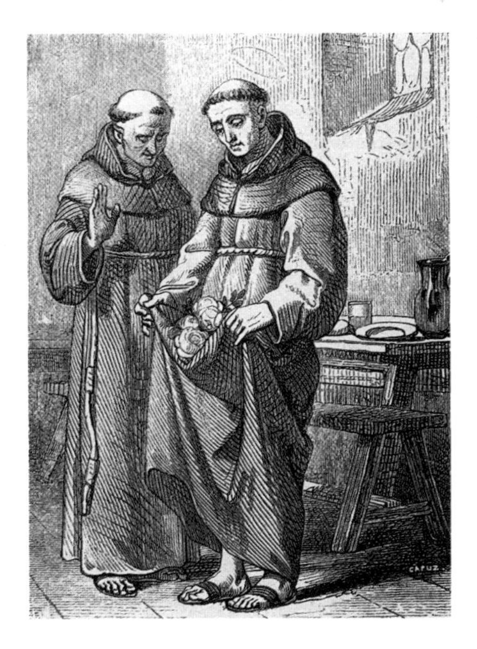

They, like the monks, had to deal with whatever ailment showed up at their door, but they often moonlighted as midwives and, occasionally, in the darker aspects of childbirth.Their kind of lore – based on treatments known to work locally, using plants found in the area – had been passed down through generations. The popular healers were not so different from some of the high-flying scholars in that they believed the body to be a delicate balance of nature, spirits and the heavens. They made full use of any form of healing practice, from soothing balms to magical amulets and even the odd spell.

As medicine spread outside the monasteries, professional (male) physicians were dismissive of folk medicine, not least because they needed to maintain their own custom. Some of the most unscrupulous even began to intimate that the agendas of such women might not be entirely pure.

Above Educated in Latin and Greek, orders of monks, such as Benedictines, placed great importance on the study and medicinal uses of plants and herbs.

Things became serious as physicians began to increase their power. With the age of discovery, new herbs, medicines and practices were moving around the world at a dizzying rate. It was time medicine saw some regulation. Serious, educated men pointed out the burgeoning breed of "quack healers" and petitioned for the power to dictate who got to call themselves a "physician". In 1518, King Henry VIII granted a Royal Charter for a new society (though Henry sat on both sides of the fence – his Herbalists' Charter of 1543 provided amateur practitioners with legal rights). The College of Physicians wouldn't add the word "royal" to its name until the end of the seventeenth century, but they were still empowered to grant licences and make rules in the meantime. They created exams for would-be practitioners, who were required to become members of the college (at a price, of course) if they passed.

The physicians were not universally loved, especially by those they accused of engaging in "malpraxis". Applicants who hadn't studied at Oxford or Cambridge universities could not become a fellow of the College and, naturally, women were not allowed at all (the first woman was awarded a licence in 1910).

However, they did do good work. The College improved standards and petitioned for public health measures, reporting on problems experienced by industrial workers in 1627 and gin drinkers in 1726.

In 1618, the College of Physicians created the *Pharmacopoeia Londinensis* ("London Pharmacopoeia"), a standardized list of medicines and their ingredients. It was the first of its kind ever published in England and was so important it was used until 1864. It had just one problem: it was in Latin, effectively stopping most people from accessing it and compounding the stranglehold the physicians had over the medical profession. The *Pharmacopoeia Londinensis* was heavily copyrighted and incredibly expensive.

They may have had a monopoly, but the College's physicians couldn't meet demand. There were usually just 60 fellows and 110 licentiates (men with a licence to practice, on payment of an annual fee) at any one time. It wasn't nearly enough. Even if the poor managed to engage a physician, they couldn't afford the prescription, to be dispensed by the Society of Apothecaries, who naturally added their own mark-up.

Nicholas Culpeper's (1616–54) grandmother taught him about medicinal plants when he was a child. William Turner's 1568 *A New Herball* was a favourite read. He was also fascinated by the stars and, by the age of 10, had read Sir Christopher Heydon's *A Defence of Judicial Astrology*.

Still interested in healing, Culpeper apprenticed himself to a London apothecary, who Culpeper taught Latin in return for learning his trade. When his master died in 1639, Culpeper continued the business. He married and set up a practice in Spitalfields, which was then a country town on the edge of the City of London.

While not particularly flush himself, Culpeper took a relaxed attitude to payment from people he knew couldn't afford his preparations. He never denied anyone treatment and often charged little or nothing for his services. He hated the College of Physicians' stranglehold, calling them "bloodsuckers, true vampires". He was so outraged at their purging, bloodletting and prescribing of medicines that were actually poisonous that he was surprised some patients recovered at all.

The English Civil War (1642–51) saw serious change for Culpeper. He was considered too valuable as a field surgeon to fight on the front line of the Parliamentarian army, which disappointed him, but he admitted "most of the barbers and physicians are royal asses".

The Commonwealth brought some good news for Culpeper: official censorship was abolished. The Stationers' Company, which had controlled printing privileges since Elizabethan times, was disabled and a flush of new books previously vetoed by them appeared on the market.

The College of Physicians had been seriously depleted by the war. While this should have been cause for Culpeper's celebration, it presented a problem – there was no one to treat the sick. His solution: an English translation of the *Pharmacopoeia* "so that all my fellow countrymen and apothecaries can understand what the Doctors write on their bills". It was about time, he argued, railing, "Hitherto they made medicine a secret conspiracy, writing prescriptions in mysterious Latin to hide ignorance and to impress upon the patient. They want to keep their book a secret."

A Physical Directory, or a Translation of the London Dispensatory was published in 1649. It was accurate, clear and complete, and included the common English name for every plant. To add insult to injury, Culpeper went one step further, annotating the work and pointing out sections he didn't agree with in rather less-than-diplomatic language. He even included cheap, wild-plant alternatives for some of the more exotic species recommended. The physicians were incandescent.

Opposite Engravings of plants including henbane, hemlock and St John's Wort, from the tome known as *Culpeper's Herbal*, originally published in 1652.

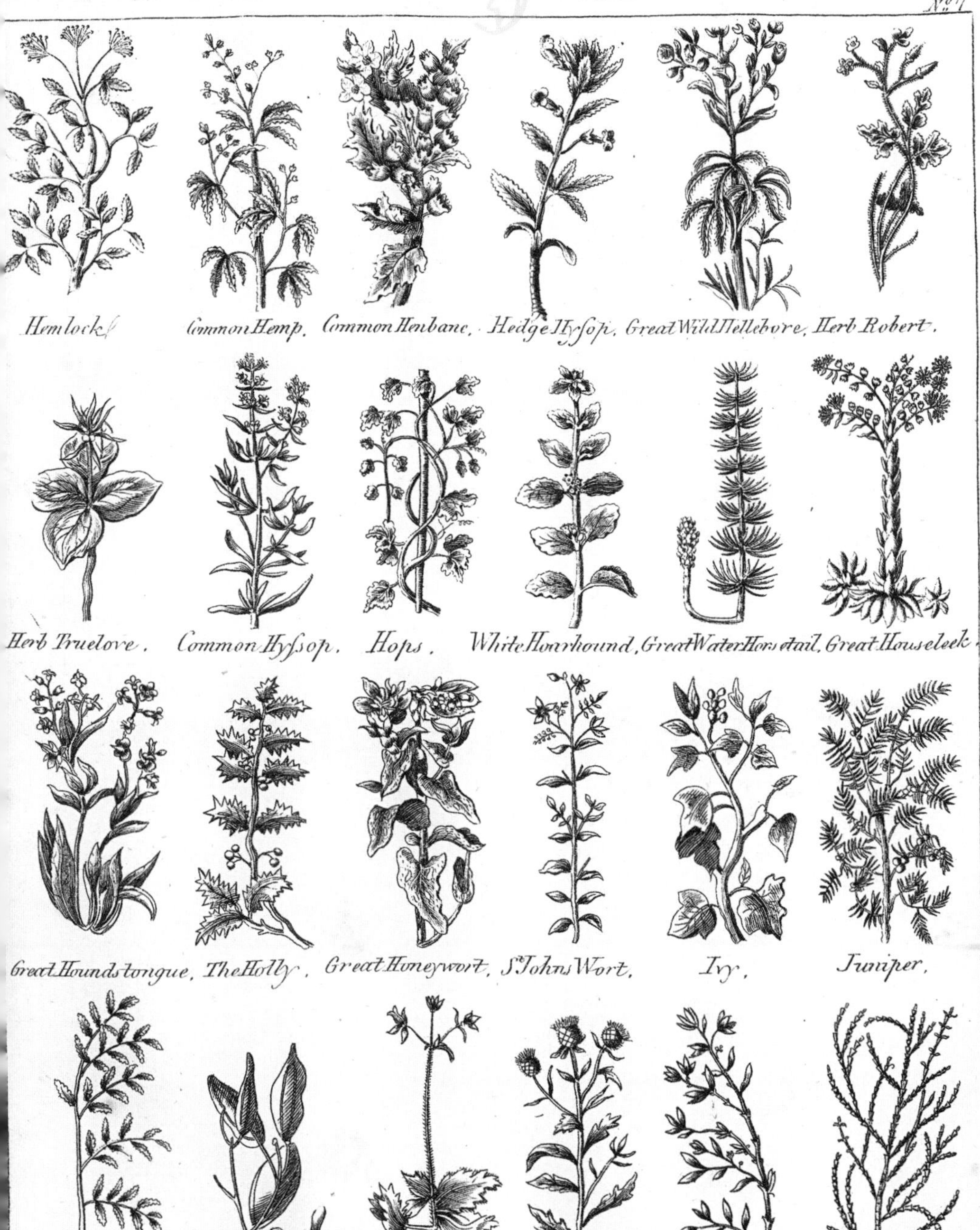
Hemlock
Common Hemp.
Common Henbane.
Hedge Hyssop.
Great Wild Hellebore.
Herb Robert.
Herb Truelove.
Common Hyssop.
Hops.
White Hoarhound.
Great Water Horsetail.
Great Houseleek.
Great Houndstongue.
The Holly.
Great Honeywort.
St. Johns Wort.
Ivy.
Juniper.
Jujube Tree
Indian Leaf.
Kidney Wort.
Common Knapweed.
Common Knot grass.
Kali.

Mr NICHOLAS CULPEPER.
DEPARTED THIS LIFE
ENGLI
H
Upwards of
MEDICINAL
The CURE of
RULES for Compounding
FAMI
And

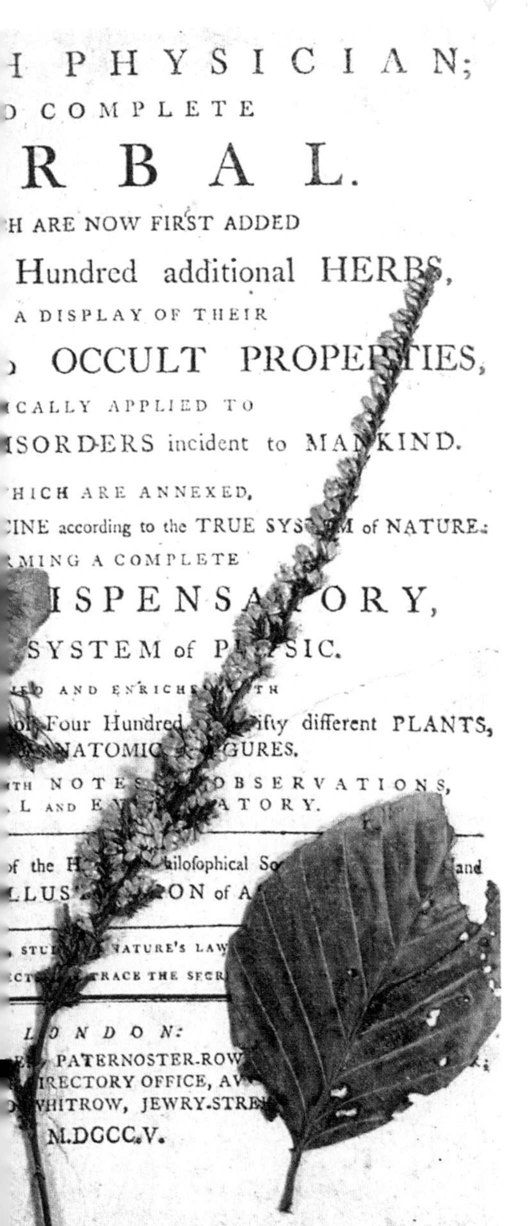
PHYSICIAN;
COMPLETE
RBAL.
H ARE NOW FIRST ADDED
Hundred additional HERBS,
A DISPLAY OF THEIR
OCCULT PROPERTIES,
CALLY APPLIED TO
ISORDERS incident to MANKIND.
HICH ARE ANNEXED,
INE according to the TRUE SYSTEM of NATURE.
MING A COMPLETE
ISPENSATORY,
SYSTEM of PHYSIC.
AND ENRICH... TH
Four Hundred ... fifty different PLANTS,
ATOMIC... GURES.
TH NOTES ... OBSERVATIONS,
L AND E... ATORY.
of the H... hilofophical S... and
LUS... ON of A
STU... NATURE'S LAW
CT... TRACE THE SECR
LONDON:
PATERNOSTER-ROW
IRECTORY OFFICE, AV
WHITROW, JEWRY-STRE
M.DCCC.V.

"The *Pharmacopoeia* was done (very filthily) into English by one Nicholas Culpeper", fumed the Royalist newpaper, *Mercurius Pragmaticus*. There was nothing they could do about it, though – their carefully written copyright wasn't worth the paper it was printed on and the people now had their own ways of getting well.

In 1652 Culpeper published the book for which he will always be remembered. He called it *The English Physician: Or an Astrologo-physical Discourse of the Vulgar Herbs of This Nation*. We know it as "Culpeper's Herbal".

It wasn't the only herbal around – William Langham's *Garden of Health* came out in the same year and Culpeper himself had clearly consulted others. What made *The English Physician* unique was that he had written it with ordinary folk in mind. Lavish woodcuts and leather bindings were not for him. This work was plain, small (which meant it was easy to carry), quick to consult and, most importantly, cheap. Costing just three pence, half an average worker's daily pay, it was soon within the pockets of the masses.

Culpeper believed English bodies needed English herbs, so he looked for common wild plants that could be easily harvested. He used the classical doctrine of signatures and his passion for the stars saw him appoint a star sign for each plant. Certain religious people took offence at this: Mr Culpeper clearly didn't know his Bible – it was quite obviously stated therein that the heavens were created *after* the plants.

The English Physician was an instant sell-out. Culpeper had to make several editions and, of course, without copyright, he was now the victim of

Left Frontispiece of an early nineteenth-century edition of Nicholas Culpeper's *The English Physician*, including a highly romanticized image of the author.

constant plagiarism. Culpeper, whose glass house wouldn't stand up to many stones on that account, made lists of the errors in the forgeries so people wouldn't suffer from wrong advice.

The work became *Culpeper's Herbal* in 1794. By then, editions were often highly illustrated and hand-tinted and had portraits of the author inside the front cover. Over the years, his likeness became less and less believable, but it's hard to imagine he would have cared.

An edition from 1810 has coloured plates of herbs and stars, and even includes anatomical drawings of the human body. By this time, *Culpeper's Herbal* was indispensable for housewives, who were expected to make up medical "simples" (herbal concoctions) for the whole family and their animals.

With the restoration of the English monarchy, order was returned. The stationers resumed their censorship and the physicians got their copyright back. However, the catmint was already out of the bag.

Two major schools of thought regarding medicine ruled the Western ancient world. The theory of the four humours explained the workings of the human body. First systemized in ancient Greece, the concept may be much older; interestingly, it has a not-dissimilar ancient cousin in traditional Chinese medicine. Humours were liquid forces within the body: blood, phlegm, black bile and yellow bile. They were associated with the four elements: air, water, earth and fire. Some even connected the four humours with the seasons.

In a healthy person, these four forces were in equilibrium. If the delicate balance was shifted, however, and either an excess or depletion of one or more humours occurred, the individual would become sick. Herbs and other medicines were prescribed to build up insufficiencies; excesses needed to be purged with laxatives or bloodletting. The ancient physicians Hippocrates and Galen were both advocates of humoral medicine and the concept was still popular in the nineteenth century.

The doctrine of signatures suggested that plants could be matched with the organs they would

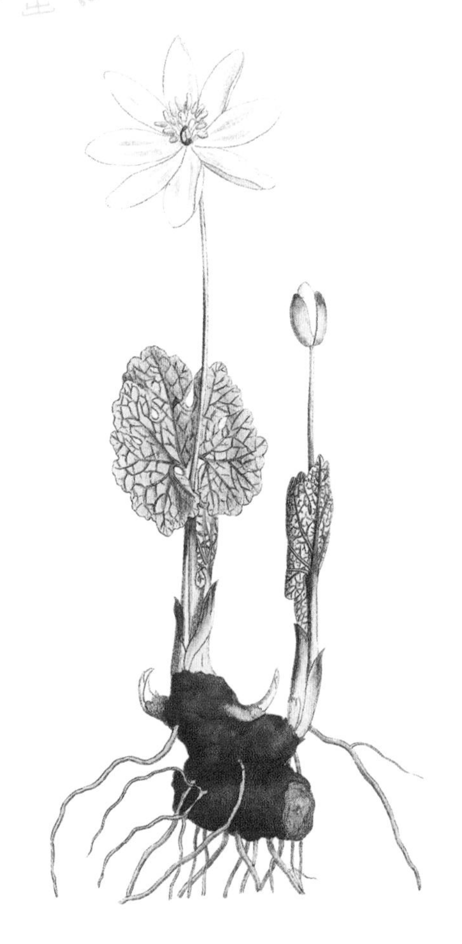

Above Bloodroot (*Sanguinaria canadensis*) from *Curtis's Botanical Magazine*, 1792.

Opposite Herbarium sheet of dittany (*Dictamnus dasycarpus*), collected in 1867.

HERBARIUM 1867. HOOKERIANUM
Ex herbario horti Petropolitani.
Dictamnus Fraxinella Pers
var dasycarpa Trautv
teste Trautv
Songarei
Schrenk

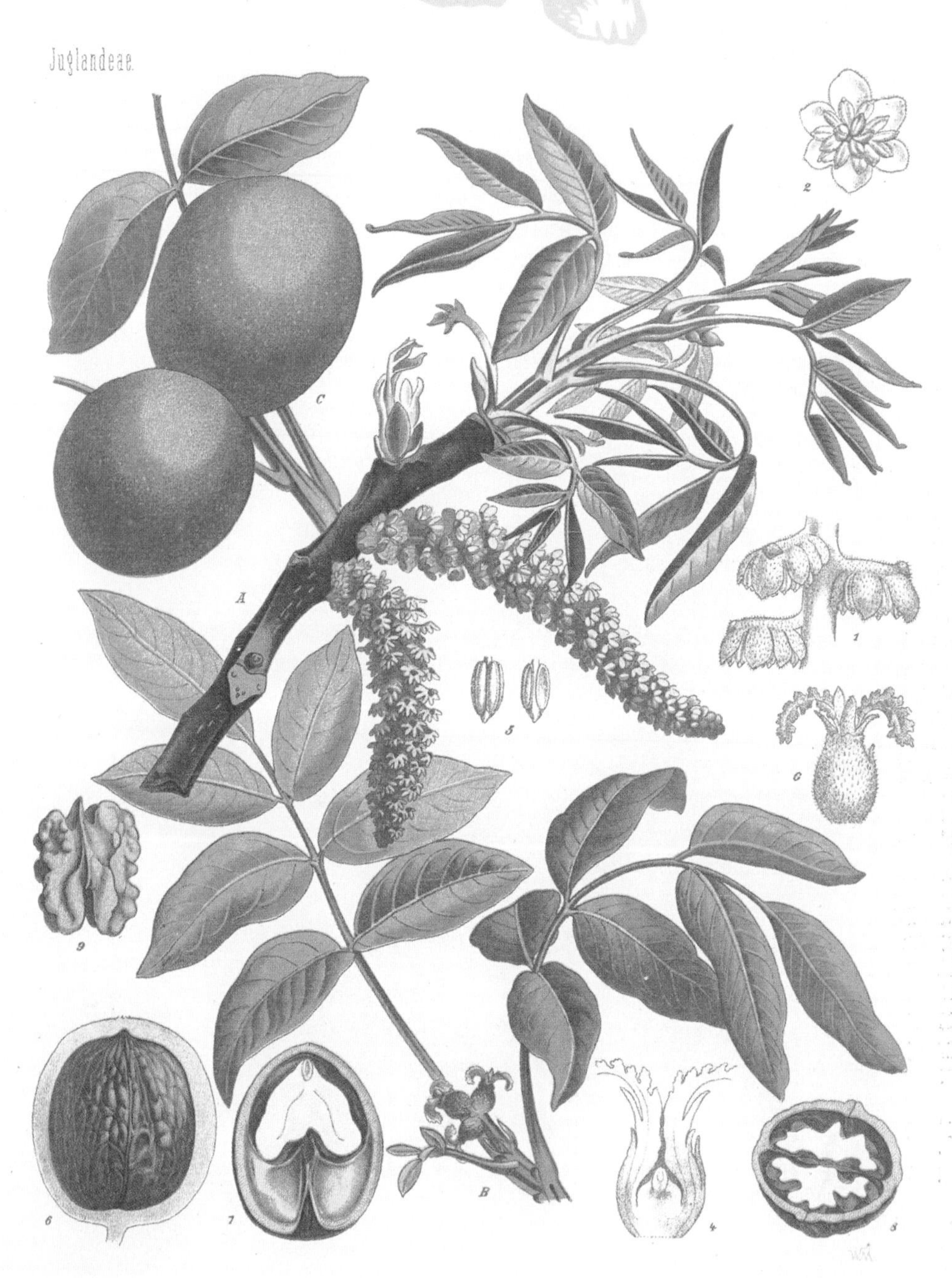

Juglans regia L.

successfully treat according to what the plants looked like. If something looked like a heart, for example, it would cure ailments on that part of the body. Bloodroot (*Sanguinaria canadensis*) secretes a red fluid, so was used for blood disorders. Walnuts (*Juglans regia*) were considered to cure headaches, while the bulbous roots of lesser celandine or pilewort (*Ficaria verna*) were believed to be ideal for haemorrhoids. This practice was popular in medieval times and got its name from the work of the Swiss physician Paracelsus, who said God had marked or "signed" each plant according to its curative benefit.

There were a lot of issues with this theory, not least that not everyone could agree what a particular plant most resembled. Sometimes treatment actually was beneficial and this has led to a theory that medieval herbalists used the system more as an aide-memoire to remember cures. For example, eyebright (*Euphrasia*) did appear to work for some eye conditions. It looked a *little* like an eye, so its physical appearance was used to remember what it was good for.

The word herbarium began to be used for private collections of dried-plant specimens in the late seventeenth century, when a botanist called Joseph Pitton de Tournefort (1656–1708) described his own samples from the natural world. Swedish botanist Carl Linnaeus (1707–78), who formalized the nomenclature of all living things, preferred the name to the previous term, *hortus siccus* ("dry garden"), and it stuck. Many of the really old herbariums are held in scientific or educational institutions. The Muséum national d'Histoire naturelle in Paris has 9,500,000 specimens; the Royal Botanic Gardens, Kew houses about 7,000,000, including 330,000 type specimens and around 95 per cent of the world's known vascular plant genera (non-vascular specimens, such as mosses, are kept at the Natural History Museum). Around 25,000 specimens are added to Kew's collection each year.

The Kew Herbarium collection dates back to William Hooker, Director of Kew from 1841, who made his own herbarium (based at his home) available to staff and visitors. It was moved to Hunter House, within the gardens, in 1852, where it was joined by another collection belonging to Dr William Arnold Bromfield. Gradually, as more collections were donated, and plants were collected by botanists on expedition, Kew's Herbarium became one of the most important in the world.

Herbariums are still vital today. The first versions were bound into book form. Modern specimens are stored separately, but they all tell a story about the history and evolution of plants. Ancient herbariums help us identify specimens from more than their names, which can change over the centuries; the dried plants remain the same, so we can tell exactly what a writer meant by a particular name at a specific moment in history. We can even check DNA to redefine family groupings and monitor modern issues, such as climate change and pollution, by comparing modern plants with pre-industrial specimens.

Opposite Walnut (*Juglans regia*) from Köhler's *Medizinal Pflanzen*, 1887. Walnuts were alleged to cure headaches.

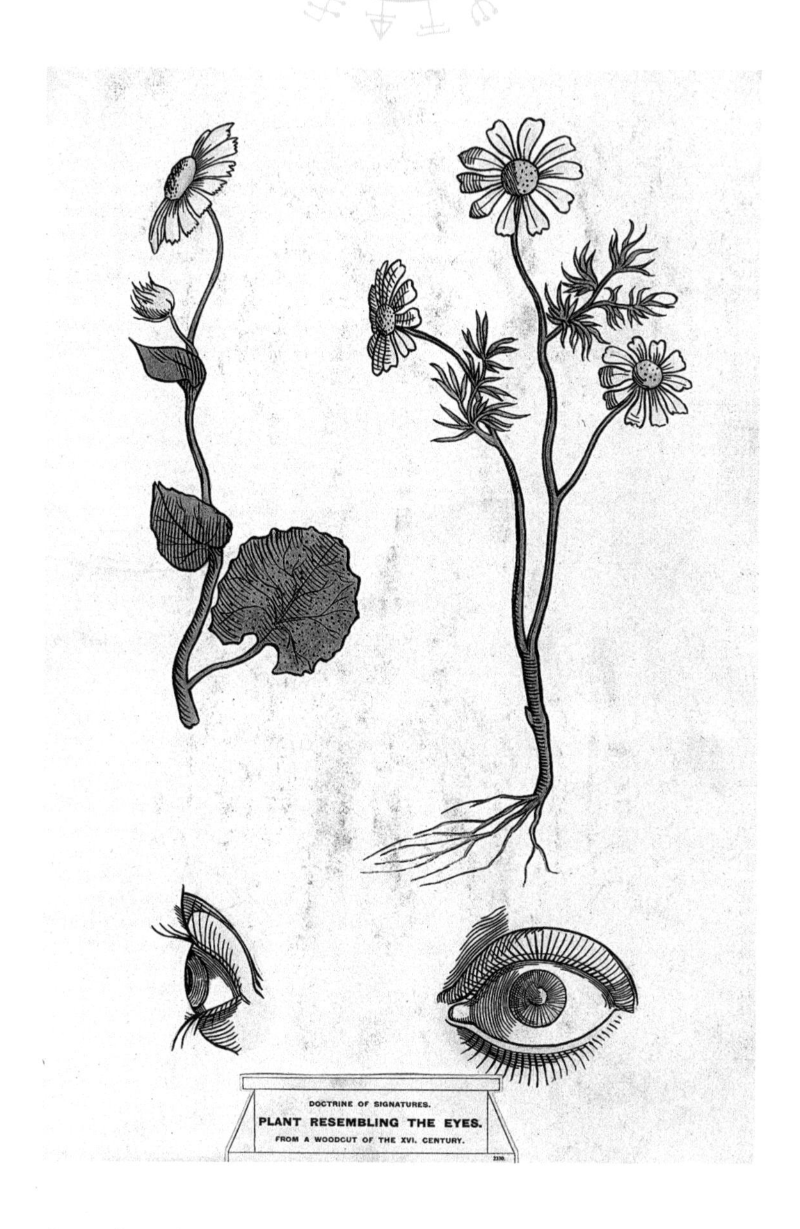

Above and opposite Two *c*. 1923 illustrations after G.B. Della Porta (*c*. 1535–1615) demonstrating the doctrine of signatures – one of flowers resembling eyes and the other a seed head resembling the human womb.

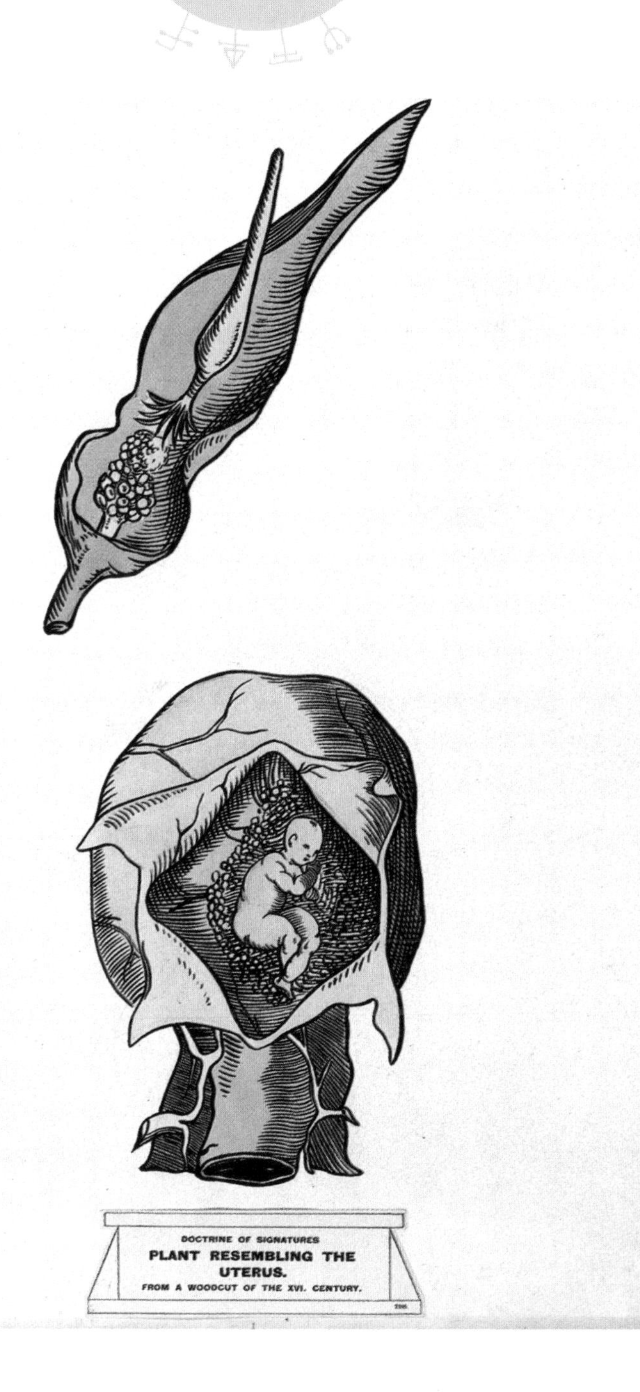
DOCTRINE OF SIGNATURES
PLANT RESEMBLING THE UTERUS.
FROM A WOODCUT OF THE XVI. CENTURY.
677

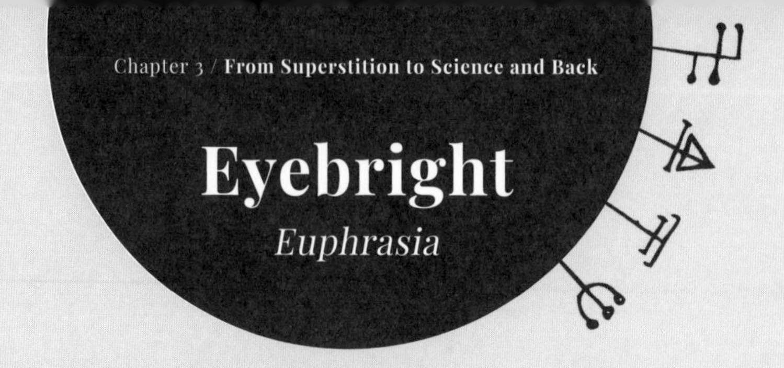

Eyebright

Euphrasia

In Scotland, if someone wants to know if their companions are telling the truth, they should carry a sprig of eyebright in their pockets.

Folklore doesn't elaborate on exactly how this works – whether the herb will force their companions not to lie, or will somehow reveal any falsehoods to the bearer. This is a common problem with relying on customs handed down through the ages: they end up being incomplete. It isn't a serious problem in this particular case, but when herbal remedies are passed down, especially by word of mouth, it's possible – indeed likely – that vital parts become lost and problems creep in.

Eyebright's qualities are well documented, not least because it is the classic example used to explain the doctrine of signatures (see page 47). There are many varieties of this common annual, also known as "fairy flax", "Christ's eyes", "bird's eye" and "joy flower", which grows on open, chalky grassland. Its jagged-edged leaves, varying between light and dark green, are easy to spot, despite hiding among surrounding grasses. Between July and September, eyebright is studded with tiny white or purplish flowers boasting the famous bright yellow "eyes" with their black "pupils".

In ancient Greece, the plant was named for Euphrosyne, the goddess of gladness and good cheer. Perhaps this is because, for centuries, it has been associated with improving eyesight. According to medieval writers, the herb could restore vision, cure inflammations and, applied to a child's eyes during a bout of the measles, prevent problems later in life. Taken as an infusion or tincture, it was also said to enhance the memory, relieve bronchial colds and dry up mucus, thanks to its powerful tannins. Gervase Markham, in his 1616 *Countrey Farme*, recommends taking a small draught of eyebright wine each day for general health.

Culpeper declared eyebright to be a flower of the sun, believing it brought light and clarity of vision. "If the herb was as much used as it is neglected it would half spoil the spectacle maker's trade," he noted. Indeed, in France the plant is known as *casse-lunette* ("spectacle-breaker"). Due to the number of plants going by very similar names, however, it was important to know what one was doing given that they were working so close to the eyes.

Eyebright has a reputation for refusing to grow in gardens unless another grass "protects" it. We now know this is because the plant is semi-parasitic, attaching itself to the roots and stems of its next-door neighbours.

Opposite Herbarium sheet of eyebright (*Euphrasia*), collected in Orkney, Scotland, 1922.

ROYAL BOTANIC GARDENS
KEW
The London Catalogue of British Plants, Tenth Edition, No. 1265.
Name Euphrasia borealis, Townsend.
(fide Dennis Lumb, who saw all these specimens on 7th April 1923).
Popular English Name Common Eye-bright.
Habitat Top of rank grassy cliffs at seashore.
Height above sea level 30 feet.
Station Lingro, Scapa Bay, Saint Ola, Mainland, Orkney.
HERB. KEW.

Labiatae.
Rosmarinus officinalis L.

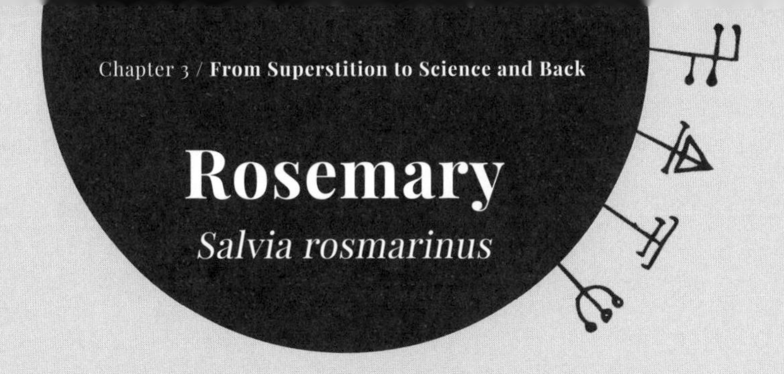

Rosemary

Salvia rosmarinus

Rosemary is another of the big-hitter medicinal herbs. First mentioned in 5,000-year-old Sumerian cuneiform tablets, it was used in all the great ancient civilizations.

Rosemary was sacred to ancient Egyptians, Romans and Greeks, and remains so for many modern-day Christians and pagans.

A small-leaved, small-flowered evergreen with leaves like pine needles, it grows wild throughout the Mediterranean region, but transplants well and became a staple of monastery gardens. Lore held that rosemary's purple-blue flower used to be white – until the Virgin Mary used the bush to dry the Holy Family's washing during their flight to Egypt and the blue from her dress stained it forever. For this reason, if caught at midnight on Twelfth Night, it would miraculously be in flower.

A gardener growing rosemary was said to never be without friends and could not be harmed by witchcraft. In the United States, if the herb was sited by the door it was considered particularly lucky. It was, however, another one of those herbs that thrived best in the garden of a dominant woman – men often deliberately avoided growing it in case their friends noticed it doing a bit too well.

Rosemary had many alleged uses. It protected against evil spirits, fairies, lightning and injury, and brought business success and love. A sprig under the bed cured nightmares while rosemary hair rinse cured dandruff. A spoon made from the wood counteracted poisoning and a rosemary snuffbox cured the plague. Girls sometimes used the herb in divination charms to dream of their future husbands.

Nicholas Culpeper deemed rosemary a plant of the sun that could ease giddiness, drowsiness and toothache and expel gas from the stomach. Rosemary oil would lubricate sinews and joints, though he warned against overuse, calling it "quick and piercing" and recommended that only a little be taken at a time.

By far the most important folk use of rosemary is for the memory, an association that dates back centuries. As Shakespeare's Ophelia says in *Hamlet*, "There's rosemary, that's for remembrance". A recent scientific study found that rosemary oil does help memory, so perhaps the tradition of giving sprigs to schoolchildren before exams is justified.

It has also become a commemorative plant. In some places, it is wrapped in white paper and placed in or on coffins while, in Australia, sprigs are worn on ANZAC Day, remembering the troops killed on the Gallipoli peninsula, where the herb grows wild.

Opposite Rosemary (*Salvia rosmarinus*) from Köhler's *Medizinal Pflanzen*, 1897.

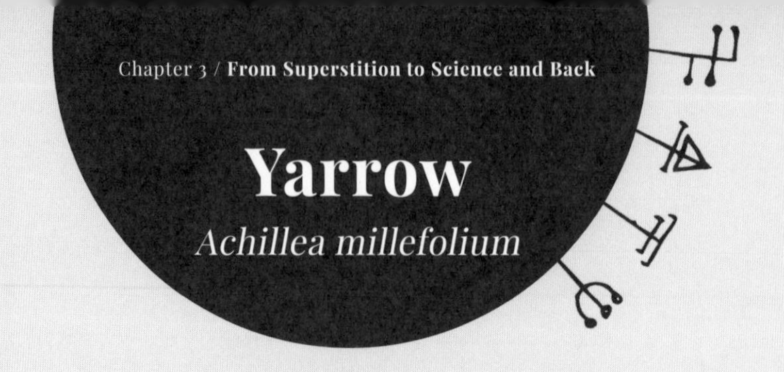

Yarrow

Achillea millefolium

The poet Homer describes how Achilles used a magical herb to staunch the blood of his army's wounds.

The Greek hero had been taught to carry the plant by Chiron the centaur, who, unlike the rest of his race, was famous for his wisdom and medical knowledge. Carl Linnaeus named the herb *Achillea* in memory of the myth. It was also formerly known as *Herba militaris*, the military herb.

Yarrow (the common name is a corruption of the Anglo-Saxon *gearwe*) continued to grace the battlefields of Europe as "knight's milfoil", "staunch weed" and "soldier's woundwort". In civilian life, it became "sneezewort" or "nosebleed", from the country tradition of stuffing the leaves up a bloody nose. Somewhat oddly, some herbals recommended it to cause bleeding rather than staunch it. Nothing is straightforward in folk medicine.

An upright plant with rough, "cornered" stems, yarrow will grow almost anywhere. The other part of its Latin name, *millefolium*, refers to its multi-segmented feathery leaves, which in the Hebrides were thought to bring second sight when applied to the eyes. The name might just as well describe yarrow's heads of hundreds of tiny white or pink flowers. Culpeper suggests boiling the flower heads whole to stop the "running of the reins" (gonorrhoea), ulcers, fistulas and, indeed, all diseases "as abound with moisture". Yarrow purportedly regulated menstruation, healed piles and lowered blood pressure. Native Americans used it for a wide variety of complaints, but the herb can cause severe allergic reactions.

Yarrow strewn on a threshold kept witches away, and eaten at a wedding would guarantee the couple would stay together for at least seven years. If they garlanded their baby's cradle with the herb, it would be protected from evil; all the better if the plant was gathered on St John's Eve (23 June). Bringing the flowers into the house, however, was bad luck.

Yarrow is an important herb of divination. Indeed, its stalks are still sometimes used when consulting the Chinese *I Ching* divination text. In Britain, it was yet another way young girls sought glimpses of their future husbands, in the usual, eye-wateringly complex ways. Some had to pull the plant from a young man's grave or pick it from a churchyard they had never seen before. In silence, some cut a sod containing yarrow to place under their pillows, while others put the plant in a right stocking tied to the left leg of the bed, then got in backward. One can only hope the resulting dreams were sweet.

Opposite Yarrow (*Achillea millefolium*) from Köhler's *Medizinal Pflanzen*, 1897.

Achillea Millefolium L.

Petroselinum sativum Hoffm.

Parsley

Petroselinum crispum

It is rumoured that only a rogue – or a witch – can successfully grow the "herb of death".

This "damned if you do, damned if you don't" legend is the least of parsley's worries. In Greek mythology, the plant sprang from the blood of the hero Archemorus after a serpent had bitten him. Roman athletes wore wreaths of parsley at funeral games and scattered the herb on graves.

Petroselinum comes from the Greek for "stone" because parsley, a Mediterranean biennial, grew wild on rocky land. Its thick, white roots, dissected leaves and small umbels of white flowers have a number of culinary and medical uses, but it came with such a list of caveats that it's a wonder anyone was tempted to grow it at all.

The bad press begins even before parsley is born. Slow to germinate, folklore holds that the seed goes down to visit the devil several times (anything between three and nine, according to who's telling the tale) before it eventually surfaces. It should be sown on a holy day or the fairies will steal it; Good Friday is the best bet. Transplanting brought bad luck, but it was even worse to receive parsley as a gift. Stealing was the only safe way of acquiring it.

Another herb that grew best in homes where the wife was dominant, parsley led some men in East Anglia to refuse to have it in the garden at all, in case they had daughters instead of sons. Harvesting was another minefield. It was said that if you cut parsley while in love, your sweetheart would die. By the same token, if you spoke a person's name while picking it, they, too, would die within seven days.

Once gathered, however, it was a useful herb. Parsley is still a major ingredient in cookery, part of the traditional French *bouquet garni*, and has been used as a breath freshener since Roman times. It was also served at Roman banquets in the mistaken belief it would prevent drunkenness. According to Culpeper, parsley aided recovery after childbirth, comforted stomach ache, eased period pain, provoked urine and "opened the body to break wind". Fried in butter, it supposedly eased breast pain and removed bruising after falls.

For all its associations with death in many cultures, within the Jewish faith, parsley is a symbol of rebirth and springtime, often used at Passover as *karpas*, one of the symbolic foods served at the festive meal known as the Seder.

Opposite Parsley (*Petroselinum crispum*) from Köhler's *Medizinal Pflanzen*, 1897.

Chapter 4: The Seasons

Even in the rush of the
modern world, most of us
pause to acknowledge the
changing of the seasons.
In days gone by, however,
knowing what each time of
the year would bring was
a matter of life or death.
Missing the best time to
plant or failing to harvest at
the correct time could see a
family starve.

In 1752, Britain changed its calendar from the old "Julian" style to the Gregorian version that most of Europe was already using. People rioted, demanding back the 11 days of their lives that had gone "missing" in the switchover, but there would be far-reaching effects too.

Any festivals set by the phases of the moon continued as ever, but set dates, such as Midsummer and Christmas, now fell on the "wrong" days. Plants didn't know this, of course, but it muddled folklore, making some customs even stranger to modern eyes.

Spring was hugely important to our ancestors. The Romans celebrated Flora, goddess of flowers; the Greeks worshipped Persephone's return from the Underworld. Days had been getting longer since the winter solstice (21 December) and people were tired of root vegetables, dried beans and withered herbs. They gathered the fresh, sweet tips of spring greens such as Good King Henry (*Blitum bonus-henricus*), dandelions (*Taraxacum officinale*) and chickweed (*Stellaria media*) for food and medicine.

Gardeners watched for native plants peeping above ground as a sign that it was time to sow their own crops. But it wasn't a foolproof method. A cold snap was known as "blackthorn winter" because frost suddenly arrived after the blackthorn (*Prunus spinosa*) had blossomed. It could damage young seedlings, though was also said to herald a good growing season. Late "onion snow" was considered a good thing, however, or at least for the allium crop. Some gardeners used birds to decide the best time to plant. The wagtail is still known as "potato-setter", "tater setter" and "potato dropper" in some areas, while in Scotland, the first swallow was a sign to get sowing.

Easter is a lunar festival, held the first Sunday after the first full moon on or after the vernal (spring) equinox, the moment when day and night are the same length. Easter was a pivotal point for gardeners. It is still the busiest weekend of the year for garden centres. Good Friday was particularly auspicious because the devil was powerless. If you sowed gillyflowers (carnations, *Dianthus caryophyllus*) at noon on Good Friday, they would miraculously bear double blooms.

Herbs of the spring were many and plentiful. The bright yellow flowers of the common gorse, with their light, coconut fragrance, meant kissing was back in season. They could also be put in streams to attract gold. Green alkanet (*Alkanna tinctoria*) flowers in the summer, but Culpeper points out that its long taproot, used for dye and numerous ointments, is at its best before the herb shoots up a stalk.

For a time of year bursting with green promise, late spring and early summer were, surprisingly, known as the "hungry gap": a time when stored supplies had dwindled and new plantings were not yet ready, but eagerly awaited.

Midsummer is still considered by many to be the most powerful time of the year, marking the beginning of three months of plenty, but there is a little confusion as to what midsummer means.

Opposite Herbarium sheet of Chickweed (*Stellaria media*), collected at Kew, 2008. The plant is one of the first herbs of spring.

The Wild Flora of Kew Gardens
Name: *Stellaria media* (L.) Vill.
Vern. name: Chickweed
Location: North Arboretum: on a small heap of soil being stored in the Paddock behind the Banks Building (zone 104)
Notes: Luxuriant shade form
Date: 23 June 2008
Collector: T.A. Cope
No.: RBG 115

Solstitium is Latin for "sun standing still". True solstices, in winter and summer, are fleeting moments when the sun reaches its northernmost or southernmost point, but the day around that point will be the very shortest or longest of the year. In the northern hemisphere, the summer solstice falls between 20 and 22 June, but one of the most significant days in the Christian calendar is St John's Day on 24 June. Add to that St John's Eve, 23 June, and the old midsummer, 6 July, and there are a lot of auspicious dates upon which to see fairies, dream of husbands, avoid the devil, drive out evil and cure various ailments.

Many herbs, including vervain (*Verbena officinalis*), sage (*Salvia apiana*), elder (*Sambucus nigra*) and, of course, St John's wort (*Hypericum*) are believed to reach the peak of their powers on the feast of St John. Counterintuitively, it is recommended to cut weeds on the full moon nearest 22 June as they will be at their weakest. Rain on St Swithin's Day, 15 July, will bless the apple orchard – though, traditionally, if it does rain that day, it will continue to rain for 39 more.

As the summer ripens, the herbalist would have looked to harvest plants while they were still potent. The daisy-like feverfew (*Tanacetum parthenium*) was believed to reduce fevers, inflammation and headaches. Legend has it that sticky, clinging goosegrass (also known as cleavers, *Galium aparine*) staunches blood and, taken in wine, relieves adder bites. The cotton thistle (*Onopordum acanthium*), which Culpeper notes may hurt the finger but help the body, was thought to be particularly good for cricks in the neck.

Opposite Sixteenth-century engraving of "storm callers" – witches summoning weather systems.

Mild nights and a full, harvest moon see the dog days of summer, heralding autumn, but there is still much to be done. Farmers have been bringing in their crops since Lammas (1 August) but it is traditionally a busy time in the still-room, too, foraging hedgerows for herbs, berries, fruits and fungi, preserving them in alcohol, syrups and drying them. The berries of Lords and Ladies, aka cuckoo pint (*Arum maculatum*), for example, were believed to ease gout when beaten with ox dung. German folklore may hold the mulberry (*Morus*) as evil because the devil cleans his boots with the roots, but its bark purportedly kills worms in the belly, its leaves relieve piles and its fruits make a syrup that will heal sores in the mouth or throat.

It is a time for cleansing, both physically and spiritually. Native American "smudging" is a smoke bath to cleanse the body and physical spaces. Bundles of herbs, including tobacco (*Nicotiana*), sage (*Salvia officinalis*), cedar (*Cedrus*), lavender (*Lavandula*) and grasses are burned and the smoke directed around the body with a feather or feather fan.

The autumn equinox falls around 21 September, a time of harvest suppers and general plenty. In days gone by, Michaelmas (St Michael's Day, 29 September) traditionally saw the last day of the harvest. The Michaelmas daisy (*Aster amellus*) is one of the last blooms of summer and is, therefore, a flower associated with farewells. By Halloween (31 October) and All Souls' Day (1 November), the wise man or woman will be prepared for days of darkness and shadow.

Traditionally, heavy crops of berries, especially holly (*Ilex aquifolium*), predict a hard winter, as do thick skins on apples and onions. This was the time to hunker down at home, using herbs in their various preserved forms to liven food and treat the diseases of the cold winter.

It wasn't all drudgery. Now that the harvest was in, in pre-industrial rural areas there was little to do and this in itself was cause for celebration. The Roman festival of Saturnalia, Norse Yule and Christian Christmas all roughly fall around the winter solstice and, after a hard year, it was time for feasting. Any plants that provided colour were fair game for party decorations, but only over the very holiest days. Holly and ivy (*Hedera*) especially should not be brought into the house until Christmas Eve and must be removed by 6 January.

Several garden plants emit a powerful scent in winter, hoping to attract the few insects around. One of the most magical winter flowering plants is witch hazel (*Hamamelis virginiana*), whose tiny pale spidery flowers punch a strong perfume into the air. The bark and leaves are astringent and were used to treat skin conditions – it is still used in skincare products today.

The days are getting longer again. Birds begin to sing and insects buzz. A new year beckons.

Above Harvesting leeks, from medieval handbook *Tacuinum Sanitatis*, the translation of eleventh-century Arabic medical treatise *Taqwīm as-Sihha*.

Opposite Carnation (*Dianthus caryophyllus*) from *Choix des plus belles fleurs*, 1824–33.

For an Ache

Take the moist dong of doves and a good half pound of feverfew, seeth them in fresh butter til they be thick to make a plaister to lay to the place where the diseased is grieved.[1]

Œillet Variété.

P. J. Redouté. Langlois.

Left Herbarium sheet of sage (*Salvia officinalis*), collected in the UK, 1880.

Opposite Herbarium sheet of ivy (*Hedera helix*), collected in Barham, UK, 1895. One of the few plants remaining green in the dark months, ivy was gathered for Christmas revels but its use in winter festivities likely goes back much further in time.

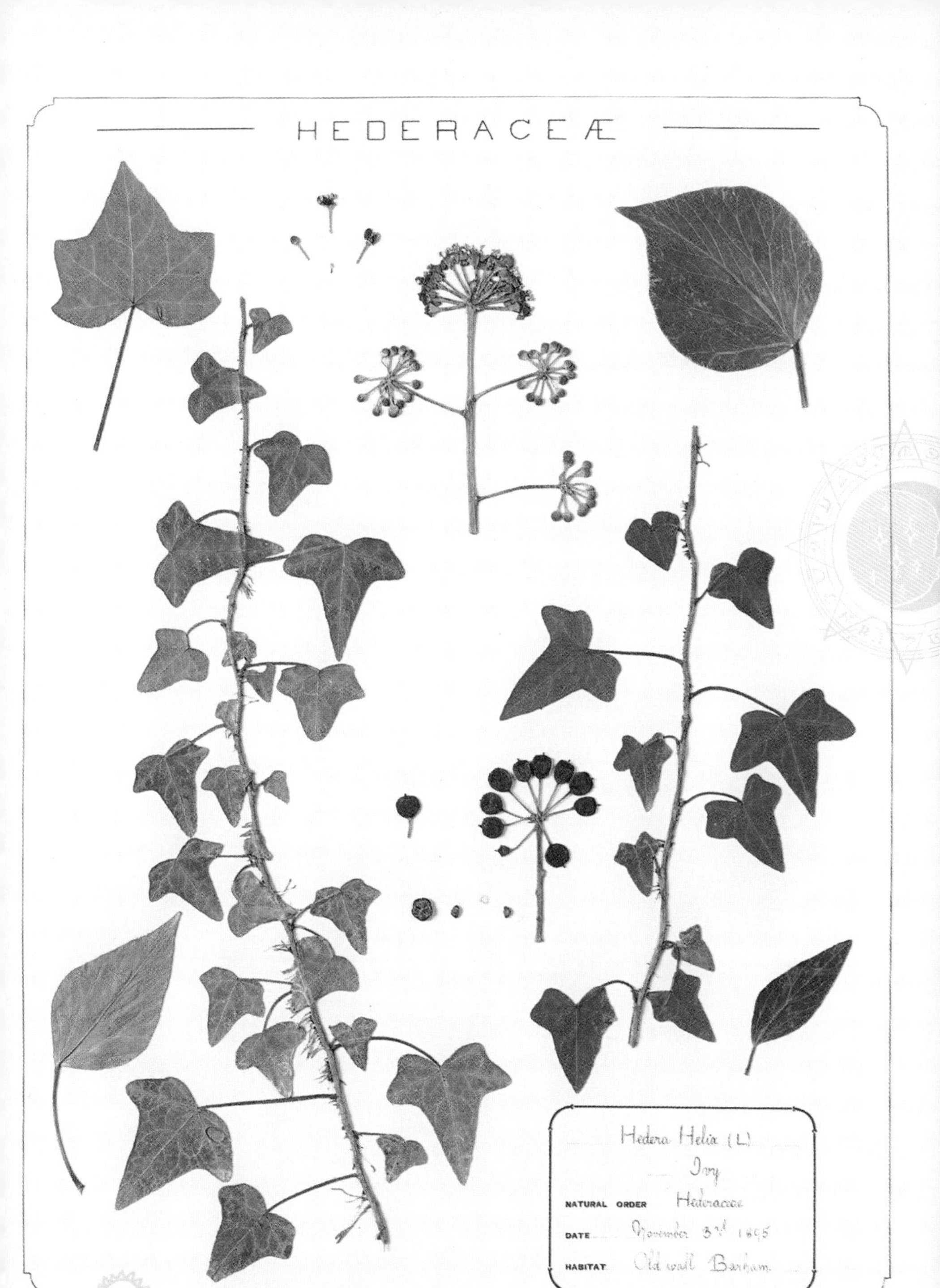
HEDERACEÆ
Hedera Helix (L)
Ivy
NATURAL ORDER Hederaceae
DATE November 3rd 1895
HABITAT Old wall Barham

Cherry

Prunus avium

Cherry blossom is one of the great symbols of springtime. Nowhere is it more important than in Japan, where the plant, known as *sakura*, takes on an almost magical quality.

Hanami, or "flower viewing", is one of Japan's most important traditions, dating back to the Tang dynasty (618–907).

The precise time when the Maiden of Spring flies across the land from south to north, awakening the sleeping trees with her warm breath, is eagerly tracked by TV weather forecasters. This is possible because, although there are many wild – and cultivated – varieties of cherry that all bloom at slightly different times, the most predominant variety, *somei-yoshino* (*Prunus* x *yedoensis*), is a clone; every plant shares the same DNA.

For the Japanese, *sakura hanami* is a time for both joy and reflection. The cherry's delicate blossoms last less than a week, reminding us of the fleeting time humans are allotted in life. It was the chosen flower of the samurai warrior class, symbolizing a short but glorious existence.

Kodama, ancient cherry trees, are inhabited by *kami*, or spirits. Each has its own legend – for example, *Uma-sakura* or "milk-nurse cherry" – contains the ghost of a wet nurse who sacrificed her own life to save a child, or *Jiy-roku-sakura*, "cherry of the sixteenth day", which blooms at the same time each year, thanks to an old samurai who gave his life to save it.

In Chinese mythology, the cherry represents immortality, not least because the legendary phoenix sleeps on a bed of cherry blossom.

In the West, cherries had a more chequered reputation. In parts of Scotland, it was a witch's tree. An old English carol sings of cherries resolving a dispute in the Holy Family: while walking in a cherry orchard, Joseph refuses to pick flowers for his wife, snapping that whoever got her with child should do so. Jesus, still in Mary's womb, asks the cherries to lower their branches and the Virgin gets her own flowers. In the Czech Republic, a charming tradition sees people bringing bare cherry branches into the house on 4 December, the Feast of St Barbara, so they will bloom for Christmas Day.

Cherries have many practical uses. Children would chew resin from the trunks like gum, while dyers used the bark for creamy-tan colours and the roots for a reddish-purple hue. The juice was used to ease coughs and bronchitis, but cherries were considered a general tonic, as is the experience of seeing them in bloom on a sunny spring day.

Opposite Woodcut print of cherry (*Prunus cerasus*) from *Traité des arbres et arbustes qui se cultivent en France en pleine terre*, Duhamel du Monceau, 1755.

Tome I. Pl. 58.

The Wild Flora of Kew Gardens
Name: Pteridium aquilinum (L.) Kuhn
Vern. name: Bracken
Location: South Arboretum: Conservation Area (zone 310)
Notes:
Date: 21 July 2008
Collector: T.A. Cope
No.: RBG 182

Ferns

Polypodiaceae

Ferns are some of the most ancient plants on the planet. Our ancestors found them extremely mysterious, both spiritually and botanically.

The term "fern" describes a number of non-flowering vascular plants (containing tissue that conducts water and minerals throughout the plant). Some date back to the Carboniferous Period, about 350 million years ago. They are often found petrified into rocks, but botanists have discovered groves of "living fossils": 500-year-old flying spider-monkey tree ferns (*Alsophila spinulosa*) in remote areas of China.

Nineteenth-century Britain was swept up in *Pteridomania* or "fern fever". Gentlefolk scoured the land with baskets and trowels seeking specimens to join exotic, foreign ferns in their back garden "stumperies" and conservatories, but this left the countryside perilously depleted. Happily, thanks to one of the fern's most enigmatic qualities, the Victorians missed the spores, which take a long time to germinate, and only collected mature specimens.

Ferns do not reproduce via flowers, which completely flummoxed early botanists. Unable to believe there really weren't any flowers, they assumed the blossom was invisible. Thanks to the doctrine of signatures, anyone who managed to catch a fern in bloom, therefore, could also acquire powers of invisibility. Some went even further, saying the finder of a fern flower would also understand birds and animals, find hidden treasure and gain the strength of 40 men.

Alternatively, they could shoot at the sun at midday on the summer solstice. If they hit the mark, it would bleed fern seed. Elsewhere, seekers placed a pile of twelve pewter plates under a fern. At midnight on St John's Day, a blue flower would bloom. The seed would pass through the first eleven plates and sit on the twelfth – easy.

Even without the invisibility they apparently conferred, ferns were useful. The curling shoots of some varieties were called fiddleheads, harvested for food and medicine – nibbling the first fiddlehead of spring prevented toothache. Dioscorides has a long list of properties of the maidenhair fern (*Adiantum capillus-veneris*), including curing baldness and dandruff. Bracken (*Pteridium aquilinum*) was used in medieval times for bedding and, medicinally, for rheumatism, blood and bladder problems. Recent research points to possible carcinogenic connections, but there is still much for science to discover about this age-old collection of plants.

Opposite Herbarium sheet of bracken (*Pteridium aquilinum*), collected in 2008 from Kew.

Oak

Quercus

The mighty oak is one of the most important plants in folklore, venerated by, among others, the Greeks, Romans, Celts, Slavs and the Teutonic nations.

It is associated with the most powerful gods: Zeus, Jupiter, Dagda, Perun and Thor. Each of these gods ruled thunder and lightning.

Oak features in many myths, new and old. Homer tells us the rustling leaves of the ancient oak at Dodona were Greece's oldest oracle. One of its branches protected Jason and his Argonauts on their voyages. Philemon and Baucis, an elderly couple in Ovid's tale, are said to have unwittingly entertained Zeus and Hermes in their humble home. Despite their poverty, they gave the disguised gods their best hospitality and were allowed to choose a reward. Devoted to one another, they asked to die at the same moment. When their time came, they were turned into two trees, a linden (*Tilia*) and an oak.

A more recent legend, that of the fugitive Prince Charles hiding from his enemies in an oak tree, is celebrated as Oak Apple Day in England on 29 May, the day of his restoration as King Charles II in 1660. The holiday has recently been revived, though anyone not wearing a sprig of oak or an oak apple (a parasitic swelling on the tree) is, happily, no longer thrashed with nettles by the local toughs.

Oak trees could be valuable even as they stood. In some places, people afflicted by ague nailed a lock of their hair to an oak, passing the fever to the tree. In Cornwall, driving a nail into the bark was a sure cure for toothache. In Wales, merely rubbing the bark in silence on Midsummer's Day ensured health for the following year. If the ash tree came into leaf before the oak, it would be a wet summer. Acorns carried in the pockets of steeplejacks and, more recently, airmen, protected the bearers from lightning.

A wand made from oak heightened the consciousness of the magician wielding it, but even mere herbalists revered the oak tree as a medicinal wonder. Supposedly, an acorn grated into milk cured diarrhoea, while water boiled with just six oak leaves rid the body of ringworm. The inner bark of an oak, mixed with powdered acorn skin, helped patients who had been vomiting blood, while the powdered acorn, taken in wine, was a diuretic and resisted poison. The distilled water of leaf buds regulated menstruation.

Oak timber was so valued for battleships that, in Tudor times, England's forests became seriously depleted. Individual specimens often acquired mythical status, such as the Major Oak in Sherwood Forest, associated for hundreds of years with the legendary outlaw Robin Hood.

Opposite Oak (*Quercus robur*) from *The North American Sylva*, 1865.

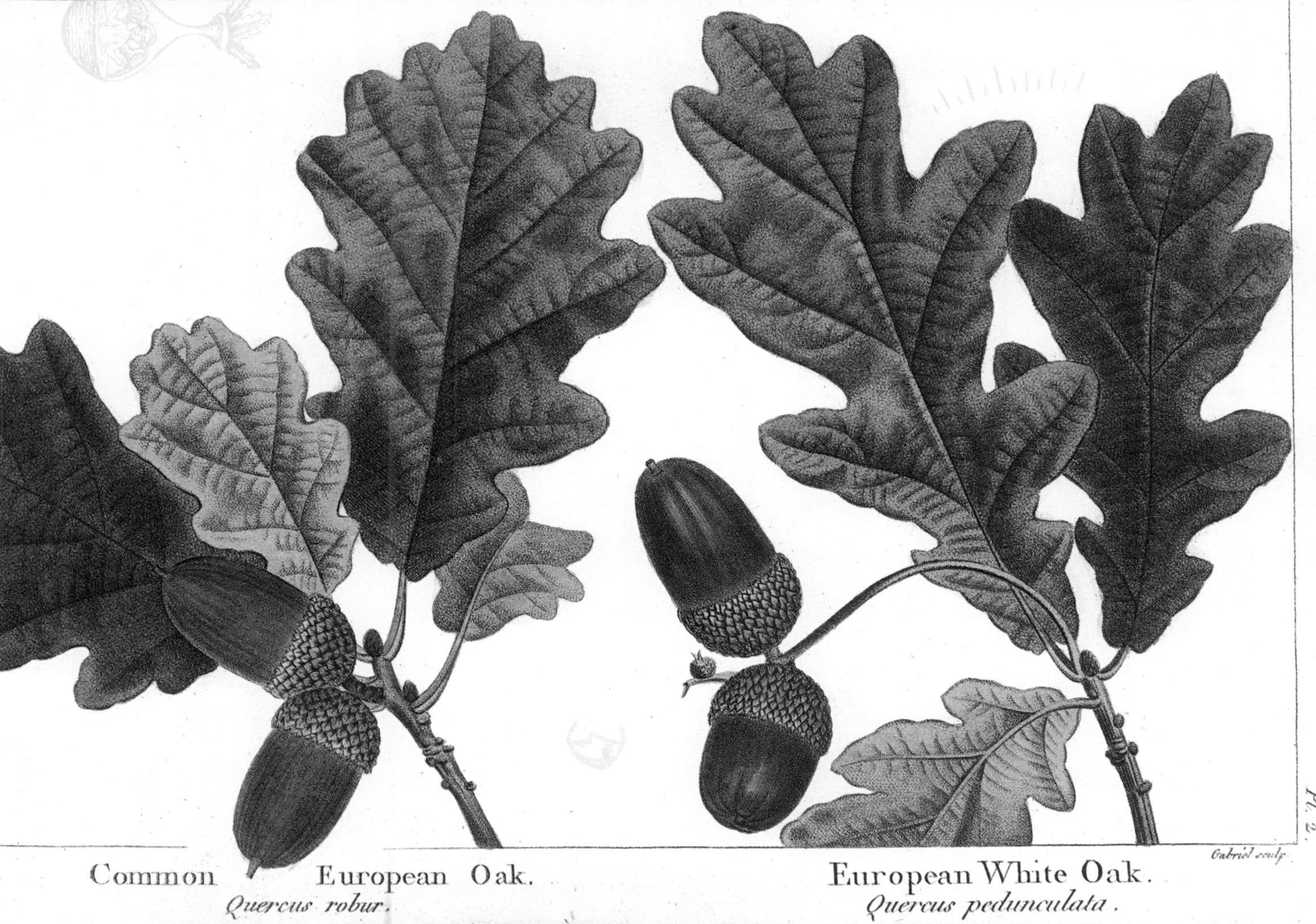

Common European Oak.
Quercus robur.

European White Oak.
Quercus pedunculata.

Holly

Ilex aquifolium

There are over 500 species of the *Ilex* genus but the most famous is the Christmas-card *Ilex aquifolium*, with its bright red berries and dark, glossy leaves.

Some know it as holm, a corruption of the Old English holen or hulver, meaning "prickle". There are vast amounts of folklore associated with this plant, often tied to the winter months.

The Romans gathered the prickly greenery for their Saturnalia celebrations and early Christians saw holly as a symbol of everlasting life. Granted evergreen status after it concealed the infant Christ from Herod's soldiers, holly originally had white berries, a privilege turned to blood red when its wood was used for the crown of thorns Jesus later wore. In Norse and Celtic mythology, the plant was linked to Thor and Taranis, gods of thunder. Holly outside the front door is still widely thought to protect a house from thunderstorms, fire and the evil eye, a centuries-old tradition mentioned by Pliny the Elder.

Holly is the tree of the gentle folk. Scottish people decorated their houses with branches to protect themselves from fairy mischief during Hogmanay, leaving an offering (usually a silver coin) at the base of the tree to appease the spirits.

Every part of the plant is toxic, but folklore still finds it indispensable. Leaves were burned like incense to strengthen magic and it was said a holly wand added protection to a practitioner while performing magic.

In Hampshire, drinking tinctures from a cup carved of holly wood would cure whooping cough, while in Derbyshire, medics literally "thrashed" chilblain sufferers with holly to let out bad blood. It was always successful; no one ever required a second treatment. Being poisonous, the plant induced vomiting and was therefore occasionally used as a somewhat dangerous purge. Culpeper suggests a holly leaf and bark poultice for broken bones and dislocated limbs.

Bringing holly into the house too soon before Christmas was bad luck, but not as bad as the luck of a manservant who failed to return home on Christmas Eve bearing a branch. The maids of the house were fully within their rights to steal his breeches and nail them to the gateposts. He didn't even get his traditional Christmas kiss.

Wearing a holly sprig to Midnight Mass risked the curse of foresight. Bearers would know which of their fellow parishioners would die in the coming year. If they failed to remove every last holly leaf from the house by Twelfth Night, it might just be them.

Opposite Herbarium sheet of holly (*Ilex aquifolium*), collected at Kew, 2009.

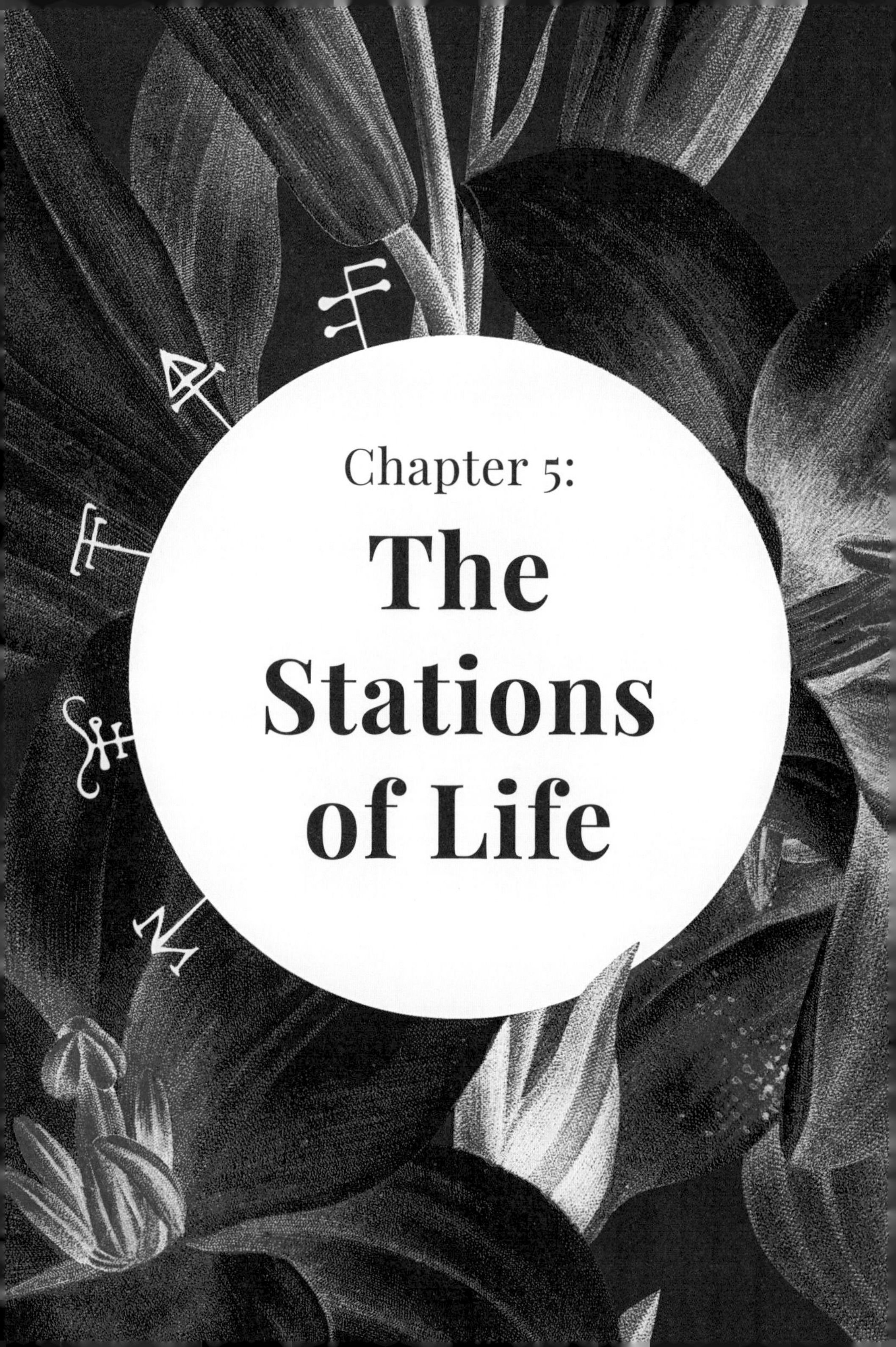

Chapter 5:

The Stations of Life

Nothing is more fundamental to the human condition than the great life phases we all share. Monarch or pauper, everyone must be born, age and die. Every civilization contrives, in some way, to mark the steps, striving to ease the physical and mental pain that tells us we're alive.

For most of the time humans have been on this planet, life was hard and short. There were many ways to die, most of them painful. Any help whatsoever, whether from divine intervention or the natural world, was welcome.

Even conceiving life in the first place was not straightforward. In many ancient civilizations, even if they'd never heard the phrase "doctrine of signatures", anything that looked remotely like the male or female genitals was worth a try in the fertility stakes: avocados (*Persea americana)*, a corruption of the Aztec name *ahuacacuauhitl*, meaning "testicle tree", due to the way the fruit often hangs in pairs; mandrake root (*Mandragora officinarum*), hung from a doorway or ceiling to encourage conception; and asparagus (*Asparagus officinalis*), which Culpeper notes "boiled in wine stirreth up bodily lust in man or woman, whatever some have written to the contrary". In Celtic traditions, phallic acorns were useful marital aids, not least because they came from the uber-male oak tree (*Quercus robur*). They were at their most potent gathered at night. Pine cones had the same effect, and pines (*Pinus*) were sometimes grown in Southern Europe as a "wedding tree" to encourage multiple offspring, insurance against the ever-present horror of child mortality. The multi-seeded pomegranate (*Punica granatum*) had obvious connections with fecundity, while Athenian brides ate quince (*Cydonia oblonga*) the night before their wedding to ensure rapid conception. Many cultures, even into early modern times, were highly suspicious of lettuce (*Lactuca sativa*), as it was thought to make both sexes barren. White dead-nettle (*Lamium album*), also known as "Adam and Eve", was considered very lucky for lovers because, held upside down, the flower's stamens look like two people in a white-canopied bed.

There were rare occasions when it was imprudent to get pregnant. The Romans were obsessed by the herb *Silphium*. They enjoyed the roots of this mysterious plant as a tasty vegetable, its flowers as perfume and its sap, known as laser, made a tasty condiment for braised flamingos, brains and parrots. But *Silphium* had another, far more potent property. Already known locally as a panacea for ailments ranging from foot corns to dog bites, *Silphium*'s sap induced menstruation – effectively aborting any foetus a woman might be carrying and turning it into a very powerful contraceptive. The plant resisted all attempts to cultivate it, so it had to be gathered wild. Cyrene, in modern-day Libya, became wealthy thanks to this tiny herb, which even appeared on Cyrenian coins. Although there were strict rules about its harvest, smuggling was rife.

Silphium disappeared in Roman times; human damage to the natural world is nothing new. Some botanists believe the herb is still out there, however, hiding in plain sight among other wild plants around Cyrene. This brings a frisson of speculation – if *Silphium* is still growing somewhere, what other

Opposite *Cydonia oblonga* (also known as *Cydonia communis*) from *Traité des arbres et arbustes que l'on cultive en France en pleine terre*, Henri Louis Duhamel du Monceau. Engraving by Mlle Janinet, 1801–1819.

T. 4. N° 56.

CYDONIA communis. **COIGNASSIER** commun.

P. J. Redouté pinx. Mlle Janinet Sculp.

HERBARIUM 1867. HOOKERIANUM.
Th. Kotschy. Pl. Pers. austr. Ed. R. F. Hohenacker. 1845.
604. Lamium Robertsonii Boiss. n. sp.
(Ex voto Th. Kotschyi dicatum Chil'archo H. Dundas Robertsonio, procuratori rerum Britannicarum in u. Buschir.)
Corolla alba.
In glareosis alpis Kuh-Daëna. D. 10. Jul. 1842.
Lamium album L.
subsp. crinitum (Montbr. et Auch.)
Mennema
det. J. MENNEMA
(Rijksherbarium, Leiden)
VI, 1978

herbs might be waiting for us to discover their mystical and curative powers?

If contraception failed, and the woman still did not want to have the child, things became rather more dangerous. Hippocratic writings prefer the repulsively named squirting cucumber (*Ecballium elaterium*) as an effective abortifacient compared to the alternative pennyroyal (*Mentha pulegium*), which ran the risk of killing the mother along with the foetus. Alas, highly toxic pennyroyal has crashed down through the ages as a theoretical way of aborting a foetus – and has been the tragic undoing of many a woman in a desperate position.

Conception was just the beginning of one of the most dangerous journeys a woman could take. Of course, she had to confirm she was pregnant in the first place – often it was impossible to tell until several months had passed, though there were some ways of checking. One ancient Egyptian pregnancy test involved the woman urinating onto two types of grain, barley (*Hordeum vulgare*) and emmer (*Triticum dicoccon*) If they sprouted, she was with child. Barley predicted a boy, emmer a girl.

During pregnancy, women drank raspberry-leaf tea (*Rubus idaeus*) to ease delivery – not to be recommended. Modern mothers might also recognize the time-honoured practice, common in several cultures, of massaging the belly with olive oil to avoid stretch marks.

Preparing for childbirth was a major event. In ancient Egypt, women would, if they could, avail themselves of the "confinement pavilion" made with papyrus columns (*Cyperus papyrus*) and decorated with health-giving vines. In coastal parts of Australia, leaves of the emu bush (*Eremophila longifolia*) were smoked to create a sterile environment.

Opposite Herbarium sheet of white dead-nettle (*Lamium album* subsp. *crinitum*), collected in Iran, 1842.

The Aztec midwife, or tlamatlquiticitl, made steam baths of aromatic plants for the mother-to-be to ease the pain. They made tea containing cioapatli, the name given to a herb that induced contractions. The sixteenth-century Spanish friar Bernardino de Sahagún was intrigued that Aztec women not only seemed to have faster births with less pain, they recovered more quickly too.

Japanese women in the Heian period (794–1185) had more than medical complications to fear. Hungry ghosts stalked the birthing chamber, hoping to devour at least one spirit. In the eleventh-century novel *The Tale of Genji*, poppy seeds are burned to exorcize evil spirits during the confinement of the hero's wife, Aoi. We are not told if these were seeds from the opium poppy (*Papaver somniferum*), but if they were, they would also have made her sleepy and relaxed.

Both mother and baby were cared for after the birth. The Greeks used mugwort (*Artemisia absinthium*), sacred to the goddess Artemis, to expel the afterbirth. Myrrh (*Commiphora myrrha*) performed a similar function, as did the chaste tree berry (*Vitex agnus-castus*), which, legend has it, helped recovery, including stimulating lactation and menstruation. Another ancient treatment, motherwort (*Leonurus cardiaca*), was still being used in Culpeper's day. He wrote that the herb "makes mothers joyful and settles the womb".

A newborn Chinese baby was, after three days, given a special bath of water bathed with locust branches and mugwort, while Australian people used a hard, bright-orange desert mushroom (*Pycnoporus species*) as a teething ring for toddlers.

The birth of a child was something to be celebrated and, in nineteenth-century Switzerland,

trees were planted upon the occasion; an apple (*Malus*) for a boy, a pear (*Pyrus communis*) for a girl. Hebrew tradition saw cedar trees (*Cedrus*) for boys and pines (*Pinus*) for girls.

As children went to school, they were sometimes given herbs to help them progress. Viper's bugloss (*Echium vulgare*) was, according to seventeenth-century diarist John Evelyn, "good for the mind", while rosemary (*Salvia rosmarinus*) and betony (*Betonica officinalis*) improved the memory. Young Greek athletes bathed in mint (*Mentha*) to strengthen their muscles, which they massaged with oil of nasturtium seed (*Tropaeolum*) after exercising.

Coming-of-age ceremonies are ways of helping a young person through one of the most difficult times of life: puberty. Acne has been treated with the astringent witch hazel (*Hamamelis*) for millennia. In Iran, they used barberry (*Berberis*) juice for acne, while the Bundjalung Aboriginal Australians were using oil from the tea tree (*Melaleuca alternifolia*) for centuries before it was "miraculously" discovered to have antiseptic healing properties in the 1920s.

The arrival of menstruation plays a pivotal part in life. Often, periods are accompanied by pain and herbalists have written endless pages on the subject. In the ancient Greek concept of the humours, pre-menstrual pain was said to be experienced by those with melancholy, or an excess of black bile in the spleen. Some of the many cures included garden angelica (*Angelica archangelica*) root, chaste tree berries, yarrow (*Achillea millefolium*), blessed thistle (*Cnicus benedictus*), feverfew (*Tanacetum parthenium*), motherwort and fennel seed (*Foeniculum vulgare*). Abdominal pain and cramps could be alleviated with, among many others, chaste tree berries, yarrow and peony root (*Paeonia officinalis*).

On the South Dakota Yankton Sioux/Ihanktonwan Oyate Reservation, girls who have had their first period are invited on a four-day female-only "moon camp" for "Brave Hearts", a name that recalls the courage of women who retrieved the dead and wounded from battlefields. The girls raise a 13-poled tepee, one support for each moon in the year. They learn to gather medicinal herbs and wild flowers and begin to learn how they are used, along with other traditional aspects of life.

Above Woodcut of fennel (*Foeniculum vulgare*) from Gerard's *Herball*, 1597. Fennel was recommended by Culpeper to increase nursing mothers' milk.

Opposite Tristan and Isolde drink a love potion, their ill-fated romance inspired art and literature throughout the ages. Wild strawberries and violets illuminate this illustration, symbolizing love.

moult grant Joye C messs trista et yseult la
eurent ensemble le bruuage amoureux

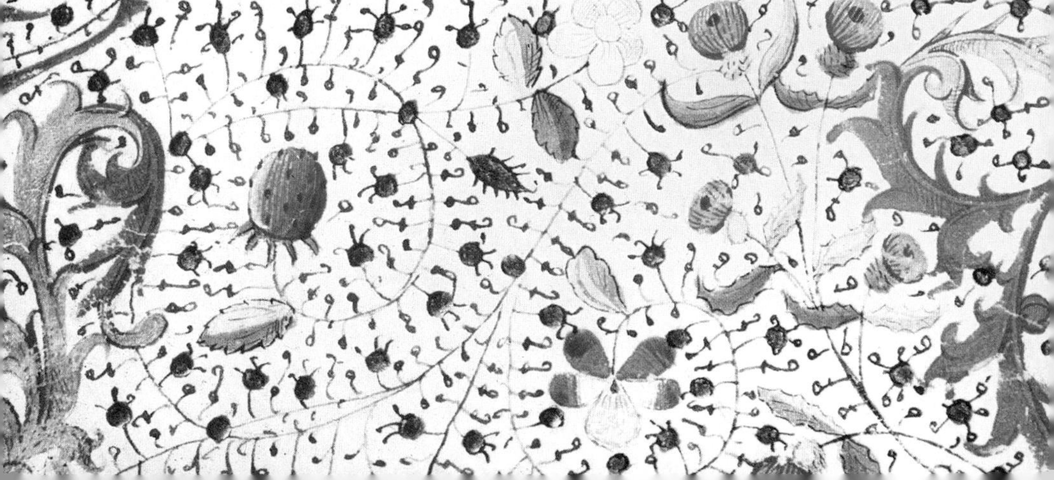

26804
EMPTY
Pinus brutia Ten., Fl. Napol. 1: lix (1811-1815)
Det. A. Farjon (RBG Kew) March 2006
possibly original material
Det. A. Farjon (RBG Kew) March 2006
HERB. J. GAY.
Presented by Dr. Hooker, February 1868.
KEW HERBARIUM
000041167

The older women bathe them in sage (*Salvia officinalis*) water and discuss sex, relationships and mental health.

Falling in love is the prerogative of the young; love herbs and potions are discussed on page 126. However, love did occasionally have an unpleasant side effect or two, and cures for venereal infections have been with us since the diseases themselves. Culpeper suggests hound's tongue (*Cynoglossum officinale*), goldenrod (*Solidago*), sarsaparilla (*Aralia nudicaulis*) and soapwort (*Saponaria officinalis*). Tragically, Rongo, traditional Māori medicine from New Zealand, didn't need such treatments until European sailors arrived. They used steam baths made from kawakawa (*Piper excelsum*) leaves.

The marriage of two young people has always been cause for celebration and the combination of brides, flowers and herbs is almost universal. Ancient Greek brides carried garlic (*Allium sativum*) in their bouquets of wild flowers to ward off evil spirits and guests held sprays of hawthorn (*Crataegus*) over the happy couple during the ceremony. Later, torches of hawthorn lit the way to the marriage chamber.

Orange blossom (*Citrus* x *sinensis*) is a symbol of chastity and has long been associated with bridal headdresses – although in France, it was only permitted to be worn by virgins. Scented vervain (*Verbena*) was often used in nineteenth-century English bouquets for its sweet smell, hidden at the back as it was considered scruffy. At the front was myrtle (*Myrtus*), carried by Queen Victoria at her wedding to Prince Albert and carried by royal brides ever since. Many Europeans throw dried rice (*Oryza sativa*) over newlyweds for good luck, but a Hindu bride throws handfuls of the grains behind her as she leaves her parents' home, to thank them for raising her. Rice symbolizes prosperity; she will also offer her groom a bowl of rice. She must not handle it, so a male relative assists, symbolizing the union of the two families. The Iranian Sofreh Aghd, or wedding table, groans with symbolic foods, including Khoncheh, seven herbs and spices that guard against the evil eye: *khashkhash* (poppy seeds), *berenj* (rice), *sabzi khoshk* (angelica), *namak* (salt) *raziyane* (nigella seeds, *Nigella sativa*), *cha'i* (black tea, *Camellia sinensis*) and *kondor* (frankincense, *Boswellia sacra*).

Few do not fear the perils of old age, and anti-ageing herbs are still prized today. The ancient Greeks believed sage warded off death, but most people needed more specific solutions. Native to the colder regions of Australia, kangaroo apple (*Solanum laciniatum*) is a member of the potato family and was used against ageing skin and uneven pigmentation. In northern Australia, the leaves and stems of the snake vine (*Hibbertia scandens*) were pounded into a poultice for arthritic joints, while in Hawaii traditional healers used *awapuhi*, or bitter ginger (*Zingiber zerumbet*) in a remedy for back pain.

Nicholas Culpeper lists many herbs useful in combatting the specifics of old age: mallow (*Malva*) and sowthistle (*Sonchus oleraceus*) for deafness, cuckoo-pint (*Arum maculatum*) and wallflower (*Erysimum*) for gout, and mugwort (*Artemisia vulgaris*) for sciatica. For wrinkles, beans (*Phaseolus vulgaris*) and cowslip (*Primula veris*).

One of the most enduring superstitions is that white flowers symbolize death. They should never be brought into the house, though some are worse than others. Arum lilies (*Zantedeschia*), even today, are funeral flowers, beautiful, elegant – and very unlucky indoors. Snowdrops (*Galanthus*) are also harbingers of doom, though a bunch in a vase on an outdoor ledge should keep the bad luck outside.

Opposite Herbarium sheet of Calabrian pine (*Pinus brutia*), collected in Italy, 1825. Evergreen pines represent longevity, peace and protection.

Across many orchards in both North America and Europe, the trees should be formally told of a family death, as should the bees in the hives nearby. Some countries dressed houseplants in mourning, while in certain areas of Germany, every pot in the house was turned out to mark the death.

In ancient Greece and Rome, rosemary and marjoram (*Origanum majorana*) were placed in the hands of the dead, and mint was used in funeral traditions because it was said to grow in the Underworld. Hebrew and early Christian traditions forbade funerary flowers because they smacked of paganism, though Christians did finally relax the rule before finding they actually liked the idea. By the nineteenth century, the famous "language of flowers" had even stretched as far as gravestones. A flower bud marked a child's headstone; if it had partially opened, the incumbent had been cut off in the prime of youth. Perhaps the best plant to have on one's tomb was a sheaf of harvested wheat, representing a well-lived life of achievement. Even the trees in graveyards were symbolic. Ancient yews (*Taxus baccata*), often older than the churchyards they stood in, protected the hallowed ground, while willows (*Salix*) wept with grief. Hawthorn represented hope, but blackberry brambles kept the devil out of graves. In Southern Europe, cypress trees (*Cupressus*) stood sentinel at gateways.

The Mexican marigold (*Tagetes erecta*) is adored as the *flor de muertos* ("flower of the dead"), dedicated to the Aztec goddess Mictecacihuatl, who guards the bones of the dead. Huge bunches of the blooms are strewn across brightly coloured altars on 2 November, *Día de Muertos*, ("Day of the Dead") representing the fragility of life as living and dead celebrate together. It is one of the happiest days of the year, bringing life's circle to a close.

To staunch a bleeding wound

Take the red tops of nettles stamped and soked wel in vinegr so lay it on the wound it will cease bleeding.[2]

Above Herbarium sheet of Mexican marigold (*Tagetes erecta*), collected in Brazil, 1950.

Opposite Mosaic of a skeleton from Pompeii, *c.* 79 CE, intended as an exhortation to enjoy life while you still can – it holds a wine jug in each hand.

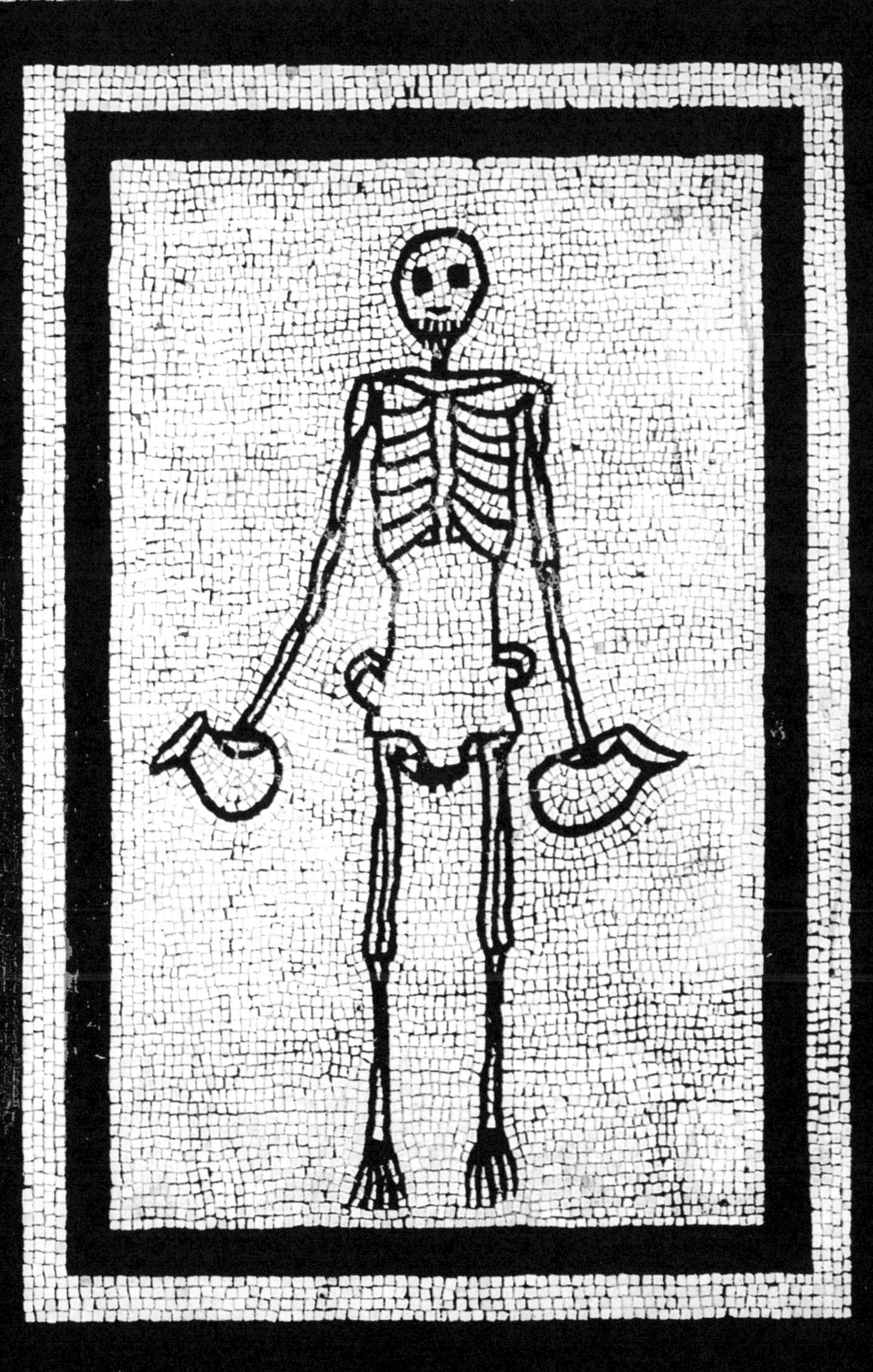

Stinging nettle

Urtica dioica

Nicholas Culpeper notes, somewhat drily, that nettles "may be found by feeling in the darkest night".

He goes on to explain how the plant consumes "phlegmatic superfluities" left by the cold and damp of wintertime.

A common countryside sight in hedgerows and woodlands, nettles were once far more important than they are now. In Denmark, archaeologists discovered a 2,800-year-old bronze urn containing the remains of a chieftain wrapped in a burial shroud woven from the finest nettle fibre. The ancient Egyptians used nettles to treat arthritis; Hippocrates writes of 61 different nettle-based remedies. Modern scientific experiments with arthritis patients have, in fact, seen the slight easing of symptoms.

Any gardener will know the adage that nettles don't sting when not in flower is nonsense, but grasping the stems firmly has the least painful result. At least there are usually dock leaves (*Rumex obtusifolius*) growing nearby, a famous folk cure for stings.

In Ireland, gathering and eating nettles at least three times during March ensured health throughout the year. In Scotland, nettles were at their most efficacious when gathered in silence at midnight.

Nettles were eaten like spinach, used to coagulate and flavour cheeses, seal leaking barrels and deter flies in the larder. During the shortages of the First World War, German uniforms were woven from nettle fibre and, in the Second World War, the military in several countries coloured camouflage equipment with nettle dye.

Nettles were used as expectorants, gargles, to provoke urine and to dissolve wind. A drink of nettle seed was thought a good antidote to dog bites, and to poisoning from hemlock, nightshade and mandrake. It wasn't. Neither was it much good at curing baldness, but that didn't stop optimistic medieval chaps dipping their combs in nettle juice and hoping for the best.

Nettles are only at their best when young and tender – thanks to the devil, who collects them to weave his shirts on May Day. After that, they're good for nothing. In the West Country of England, 2 May was "Nettle Day", when schoolchildren chased each other with handfuls of "naughty man's plaything" for the literal hell of it. More innocently, later in the year, they also made whistles from the hollow, dried stems.

Opposite Illustration of white dead-nettle (*Lamium album*) from John Curtis, *British Entomology*, 1823–40.

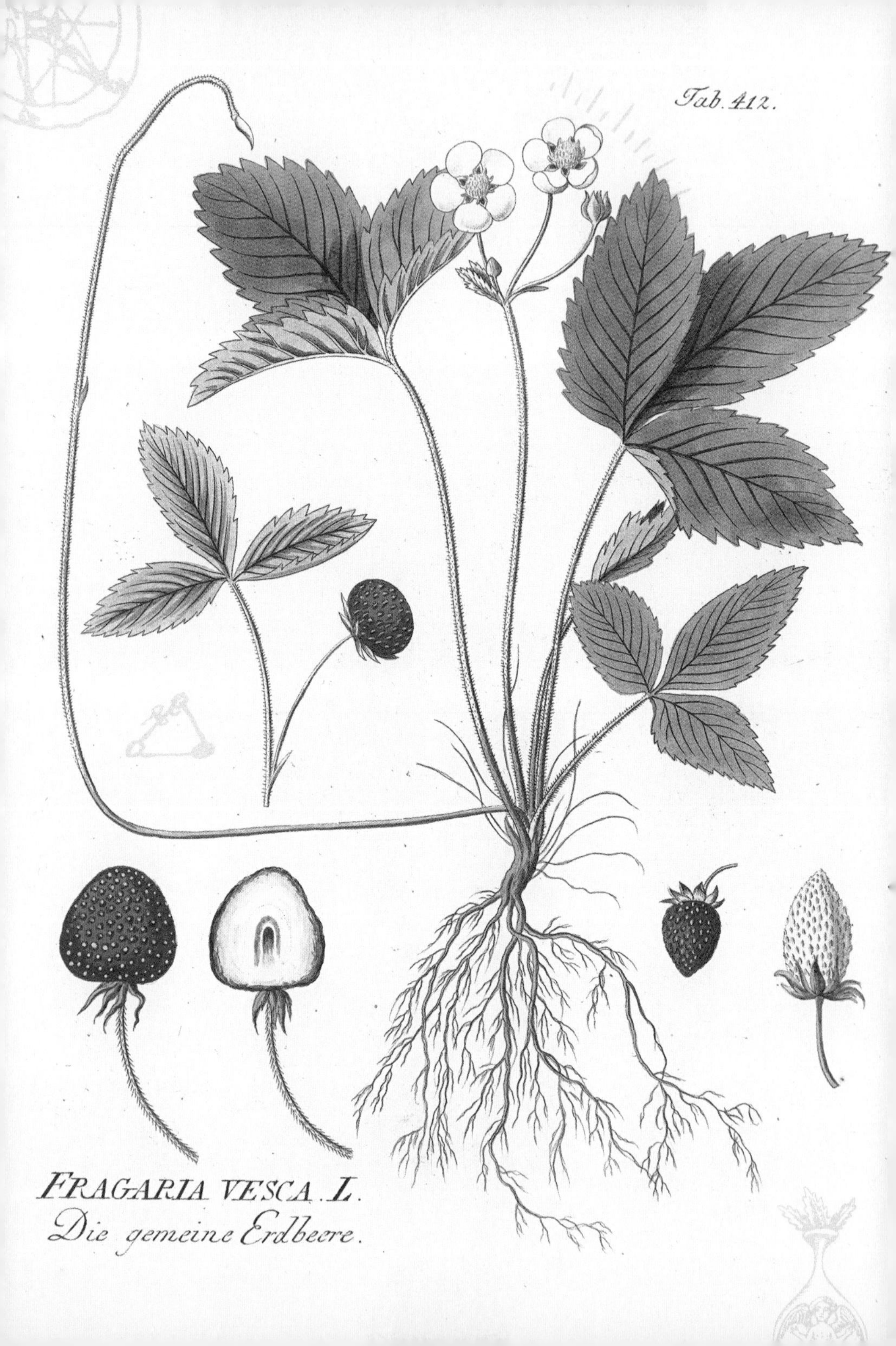
Tab. 412.
FRAGARIA VESCA. L.
Die gemeine Erdbeere.

Wild strawberry

Fragaria vesca

Highly perfumed, highly flavoursome and no bigger than a little fingernail, the wild strawberry is a long shout from the flashy, bloated monsters in the supermarket aisle.

Yet as a symbol of love, it has rarely been paralleled. Peeping shyly from dense undergrowth – just finding it becomes a delicious challenge for the would-be swain.

The Roman poet Virgil called strawberries "the children of the earth", though he warned real children to beware of serpents while gathering them. Ovid, Pliny the Elder and the Roman senator Cato all discussed strawberries, but it would take until medieval times for the fruit to reach its full, passionate potential. In vain, the Christian Church designated strawberries as the fruit of the Virgin, pointing out the tripartite leaves (symbols of the Holy Trinity), the pure-white, five-petalled flowers (Christ's five wounds) and the blood-red fruit representing drops of His blood. That voluptuous, heart-shaped berry set pulses racing ...

Strawberries were served to newly-weds with cream and sky-blue, star-shaped borage flowers (*Borago officinalis*). At Kenilworth Castle, Sir Robert Dudley, Earl of Leicester, filled his famous garden with strawberries in the hope of snaring Queen Elizabeth I's hand. The strawberry cordial preferred by a later rival, Sir Walter Raleigh, involved a gallon of strawberries in a pint of sugar-sweetened aqua vitae.

It wasn't just in Europe that strawberries were the food of love. A Cherokee legend talks of a quarrel between the first man and woman, calmed only after the Creator sent down some strawberries. The pair were duly reconciled and the human race survived.

Culpeper tells us the strawberry is a fruit of Venus, but he has little time for aphrodisiacs. Instead, he suggests bathing inflamed eyes with strawberry juice or water and boiling the roots and leaves in wine to cool the liver, spleen and blood. Every herbal had its own suggestions and parts of the body to be cured. Folklore added more. The acid in cut strawberries could whiten teeth and remove freckles. The leaves, being slightly astringent, allegedly made good gargles for mouth ulcers and poultices for wounds. Added to baths, they relieved hip pain.

William Coles, *in The Art of Simpling* (1656), suggests growing borage next to strawberries to encourage larger fruit. But by then, hybridization with new, North American strains was already underway. The fruit of love would soon be redder, bigger and showier, but it would never be sweeter.

Opposite Wild strawberry (*Fragaria vesca*) from *Icones Plantarum Medicinalium*, Joseph Jacob von Pleck, 1792

Liquorice

Glycyrrhiza glabra

The name "*Glycyrrhiza*" comes from the Greek for "sweet root" and indeed, the rhizomes of the liquorice plant contain glycyrrhizin, a substance 50 times sweeter than sugar.

This was not lost on the generations of schoolchildren who chewed the woody roots to extract every last bit of flavour. Alexander the Great and Julius Caesar's armies both chewed liquorice and Napoleon Bonaparte ate so much of it his teeth turned black. Surprisingly, the roots have also been used as toothbrushes.

A feathery, airy plant with bright green leaves a little like those of the young ash tree, and with spires of purple flowers, it is a member of the pea family (Leguminosae/Fabaceae). It can reach 1.5 to 2 metres in height and a 1-metre spread in a single season. It anchors itself with a tap root and sends out runner-like rhizomes, which become the prized, aniseed-flavoured liquorice. Growing wild on the riverbanks of south-eastern Europe and south-western Asia, it thrives in many climates, from Scandinavia to Spain.

It's first mentioned about 4,000 years ago in the Babylonian *Code of Hammurabi*, and was used in ancient Chinese and Hindu medicines. Large amounts of liquorice were discovered in the tomb of Tutankhamun. Theophrastus talks of the "Scythian root" as being useful for asthma, coughs and chest pain, while Dioscorides recommends liquorice for harshness of the throat.

Liquorice may have been brought to Britain by returning crusaders. It became so popular that, in 1305, Edward I started taxing imports of it to repair London Bridge. The answer was to grow it in Britain. Liquorice arrived in Yorkshire via Cluniac monks, and in 1760, George Dunhill added sugar to a monastic recipe to make "pomfrets" or pastilles, which would eventually become the famous Pontefract cakes. By the middle of the twentieth century, Yorkshire's liquorice "garths" covered miles of land. They are all but gone now, though some are being reintroduced.

Nicholas Culpeper loved liquorice, declaring that it could hardly be said to be an improper ingredient in any composition, of whatever intention. It relieved coughs, wheezing, shortness of breath, chest and lung pain, and cooled fevers and loosened the bowels in children. Actually, depending on the dose, it can loosen the bowels in everyone and is best taken in moderation, especially by people with high blood pressure. Nevertheless, the plant is taken seriously by modern-day scientists, too, who are interested in its potential medicinal properties. Perhaps Napoleon was right after all.

Opposite Liquorice (*Glycyrrhiza glabra*) from Köhler's *Medizinal Pflanzen*, 1897.

Glycyrrhiza glabra L.

L.
Lilium album flore erecto, et vulgare.
Weiſſe Lilien.

Lily

Lilium

Lilies come in many forms, from the charming white or purple martagon (*Lilium martagon*) to exotics from the East, such as the giant Himalayan lily (*Cardiocrinum giganteum*).

The classic meadow or Madonna lily (*Lilium candidum*) is a curious flower, a symbol of virginity, purity and innocence – and death – found at both weddings and funerals.

The Madonna lily became associated with the Virgin Mary in the Christian Church; the archangel Gabriel often carries a spray of them in Renaissance depictions of the Annunciation. It will only grow, it is said, where the mistress is master. If parents suspected their daughters of impurity they were advised to feed them powdered lily. If the girls were chaste, they would immediately need to pass water. If a man stepped on a lily his family would lose its innocence, though lilies did keep ghosts away from the garden and to dream of them brought good luck.

Forty Madonna lily petals steeped in brandy would cure boils, while the roots, roasted and combined with oil of roses, erased wrinkles and whitened the skin. Culpeper, never a man of romance, mixed the roasted root with hog's grease for a poultice to ripen and burst plague sores. It would also, he claimed, reunite cut sinews, ease burns and reduce swellings on the "privies".

Girls were warned that smelling lilies would cause freckles – a small worry in comparison to the woes of cat owners. Consuming even small amounts of lily pollen, leaves, petals or even the water they stand in is extremely poisonous to felines.

The ancient Egyptians prized lilies so much that they buried their dead with them. Greek mythology tells of how the infant Heracles sucked too strongly at Hera's breast. Most of the spilled milk became the Milky Way, but some drops fell to earth as lilies. Roman mythology has an even odder tale: that the goddess Venus was so jealous of the lily she caused its central spike to grow longer, making it less attractive. This was probably the elegant arum lily (*Zantedeschia aethiopica*).

In pagan folklore, that male, phallic spike or "spadix" is half the reason the arum lily is a symbol of sexual harmony. The other half – the white outer bract, or "spathe" – represents the female organs. However, the arum lily is not a true *Lilium*, being part of the Araceae family, and therefore closer to the woodland lords-and-ladies (see page 110). If the regular lily is associated with death, the arum is positively funereal – though oddly, it is often used in church decorations at Easter time. Even today, many people refuse to have them in the house.

Opposite Lily (*Lilium*) from *Thesaurus rei herbariae*, Georg Wolfgang Knorr, *c.* 1772.

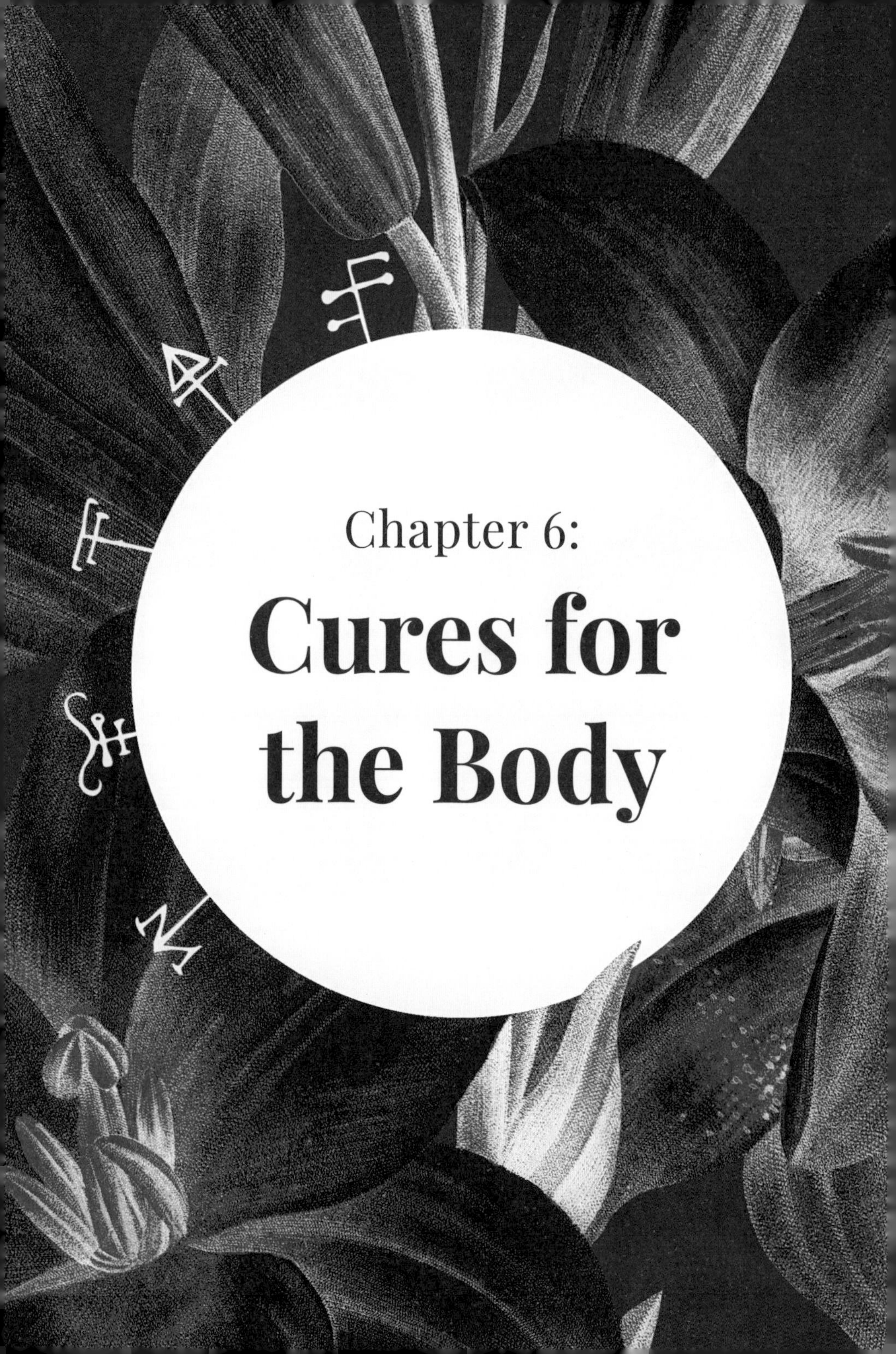

Chapter 6:

Cures for the Body

From a simple soft leaf
wrapped around a battle wound
to complex recipes handed
down from practitioner to
practitioner and accompanied
by spells, prayers and rituals,
the idea of healing herbs has
been with us since
humans evolved.

Whatever the civilization, however far back in history – or even prehistory – the same complaints crop up again and again: warts, burns, ringworm, bone ache, boils, sore throats, toothache, bad breath, bunions, flatulence and bellyache.

These are the diseases of humanity and it's only relatively recently – since the discovery of antibiotics – that we have been anywhere close to treating them. Many of the herbal remedies known to Celts (or Romans, or Greeks, or Native Americans) were still in common use in the early twentieth century. Some still are. Traditional Chinese medicine and Ayurveda are both widely practised; at least one guiding principle of both – that of custom-building treatments to the specific needs of an individual – is gaining traction in "new" twenty-first-century Western medicine.

Deciding whether to treat the whole patient, the afflicted organ or limb or the disease affecting it was of great interest to the ancients. If a leaf looked like a hand, would it cure anything that harmed the appendage, from a broken finger to a gangrenous palm? If the patient was melancholic, should they be given the same treatment for that broken finger as the phlegmatic person next to them, even though the injuries looked the same? How should dosage be decided? Too much could kill a patient, too little might not work. Should amounts be different for children and full-grown warriors? What if a poison was introduced in such tiny amounts that the body could kill it? Might it then recognize the same poison in the future and be able to destroy larger amounts of it? Could someone take a poison in small amounts every day until they were immune? In a world that had no rules, someone had to make some, and it's hardly surprising the great "fathers" of medicine, such as Hippocrates and Dioscorides, held sway for so many centuries. Few want to experiment when they're sick. Much the same thing happened in folk medicine; the trial and error of ancestors was handed down by word of mouth, then codified by the writers of herbals, such as John Gerard and Nicholas Culpeper. With the slow migration of country folk to the cities, much of the old "lore" was slowly being lost, even in mid-seventeenth-century Commonwealth England; Culpeper's herbal was one of the few cheap, reasonably reliable methods of getting well.

Choosing a treatment was a skilled job, often a curious balance of science and mystery. Nevertheless, some herbs have continued through the millennia as being reputedly "good for" certain conditions. Perhaps the most widespread of the great conditions is the most mundane-sounding: "general pain".

Opposite Fifteenth-century woodcut of apothecaries in an apothecary's shop. An apprentice works with a pestle and mortar in the background.

Chrysanthemum Parthenium, Pers.
Revision Afric. Compositae
Determinavit: J. Hutchinson
HERBARIUM 1867
120 Ex Herb. Harvey.
HERBARIUM 1867
40 Ex Herb. Harvey.

Feverfew (*Tanacetum parthenium*) has been used as a painkiller since ancient Egypt and Greece, where it got its second name, *parthenium*, because legend held it was used to cure someone who had fallen from the Parthenon temple in Athens. Feverfew was cultivated in the great Benedictine monasteries from the tenth century and gradually made its way to medieval cottage gardens. A plant that was both quick to grow and quick to apply, it was used for everything from menstrual pain and fevers to melancholy. It was also said to treat "elf-shot", where invisible fairies fired invisible arrows to humans and animals that had offended them, causing sudden, intense pain. Sometimes people even found the offending "elf arrows" (Neolithic flint arrowheads), but it's now thought the condition could have been muscle stitches or cramps. Oddly, although feverfew was known to relieve headaches, serious attention was only paid to this aspect of the herb relatively recently. It is, arguably, more used for migraines today than in the past.

The ancient Greeks knew all about the soothing aspects of opium poppy (*Papaver somniferum*) but they didn't know how to extract the most powerful (and dangerous) chemical compounds, such as morphine, codeine and thebaine, from the raw juice. Opium would have been taken in its simple form, poppy "juice" – latex extracted, unusually, from the plant's seed-head. Mint (*Mentha*) was also widely used to refresh the mind during a migraine, but that too would not have been a dramatic cure. Ancient civilizations were experimenting with willow (*Salix*) as a general cure as far back as 3000 BCE, but in the nineteenth century, a chemical from the bark was isolated, which inspired the development of aspirin.

Opposite Herbarium sheet of feverfew (*Tanacetum parthenium*), collected in South Africa, 1867.

The ancients knew the heart had something to do with pumping blood, but until the work of British anatomist William Harvey (1578–1657), they did not properly understand circulation. Blood was a magical life force to be feared and revered. Roman centurions drank the blood of their opponents; the Egyptians bathed in blood to recover from illness, and the Greeks thought excesses of blood caused certain diseases. The process of bloodletting is even depicted on ancient Greek vases.

Too little blood caused anaemia, something that could be cured by using plants that "bled", such as beetroot. Curiously, this classic of the doctrine of signatures could actually help, since beetroot contains decent amounts of iron. In traditional Chinese medicine, the heart itself was associated with the "spirit", to be treated with respect, but within a rounded view of the entire body. Traumatic experiences and lack of joy could harden the heart's arteries just as much as lack of exercise or the wrong diet.

Motherwort's benefits to the heart are reflected in its Latin name, *Leonurus cardiaca*. It's not mentioned in the works of Dioscorides or Theophrastus, but it became hugely popular in the seventeenth century, possibly due to influence from the East, where it had been used for centuries. Motherwort had a naturally sedative effect, which calmed the patient and helped "trembling of the heart". Despite its mildly unpleasant odour, by the 1600s, it was being used for palpitations, hypertension and heart irregularities as well as a host of other complaints. Gerard's lists of the conditions eased by the herb include convulsions, cramps, worms in the belly and "travails with child". Culpeper noted, "there is no better herb to take melancholy vapours from the heart, to strengthen it and make a merry, cheerful and blithe soul than this herb". Folklorist Margaret Baker relays an anonymous (and not overly catchy) saying of the time: "Drink motherwort and live to be a source of continuous astonishment and grief to waiting heirs."

In the days before refrigeration and pasteurization, diseases of the digestive system were sometimes down to simple food poisoning. Sometimes, of course, food poisoning was deliberate or "wilfully accidental", due to food adulteration – for example, unscrupulous flour salesmen bulking out their goods with chalk or plaster, or butchers' revitalizing of meat with cochineal dye to make it look fresh. Knowledge of emetics – herbs that cause vomiting – was useful to get rid of the offending food (if it hadn't already caused retching).

Ipecacuanha (*Carapichea ipecacuanha*), known as the South American "vomiting root", is a flowering plant first brought to Europe from Brazil in 1649 as a cure for dysentery. It caught the attention of French physician Helvetius, who bought every last ounce of it he could find, then created his own secret concoction. It would seem the other herbs he used were of no particular value, merely added to make his potion "special". Helvetius's secret formula cured King Louis XIV's ailing son of dysentery, and his own purse from ever being empty again. Culpeper knew the herb but, ever mindful that it was out of the financial reach of most of his readers, recommended orach (*Atriplex patula*) instead, saying that although it wasn't as strong as ipecacuanha, it didn't bind the bowels afterward.

Anyone reading Culpeper cover to cover will note the sheer number of herbs that "provoke" urine, stop urine, or treat "bloody" urine. Bladder problems were both serious and excruciating, and people were desperate to find ways of relieving themselves. The Greeks knew that urine is made in the kidneys and flows via the bladder. They were the first to "cut for stones" – making an incision into the bladder to remove accumulated hard masses of minerals. Culpeper lists well over 60 herbs that could help break bladder stones, hopefully avoiding such surgery.

To most modern gardeners, horsetail (*Equisetum*) is a pernicious weed, fit only for the bonfire. Our ancestors, however, found many uses for this prehistoric-looking plant, with its feathery yet coarse-to-the-touch leaves. Toxic to livestock and humans, it was nevertheless said to stay the fluxes, solder fresh wounds and, taken in wine, provoke urine and break up stones. At the very least, its abrasive surface made a great pan scourer.

Today's gardeners know comfrey (*Symphytum officinale*) as a superb but evil-smelling general fertilizer. Yesterday's husbandmen knew it as a powerful herb to be taken both internally, for diarrhoea, and externally, smeared on leather, as a poultice for ulcers, gout, arthritis, nappy rash, cuts and bruises. It was also said to mend broken bones, hence a variety of local names including "knit-bone" and "boneset". Recent research has, alas, cast some doubt over the safety of using comfrey. No issues have been reported over comfrey's use as a plant tonic, but wise gardeners prescribe a peg for the nose for that stench.

Cancer is as old as humanity, found in the mummies of ancient Egypt, including the world's oldest recorded incidence of breast cancer, from

Opposite Fifteenth-century engraving of human anatomy. Medieval scholars named the parts of the body and the herbs that cured them in Latin, while regular people made do with common names.

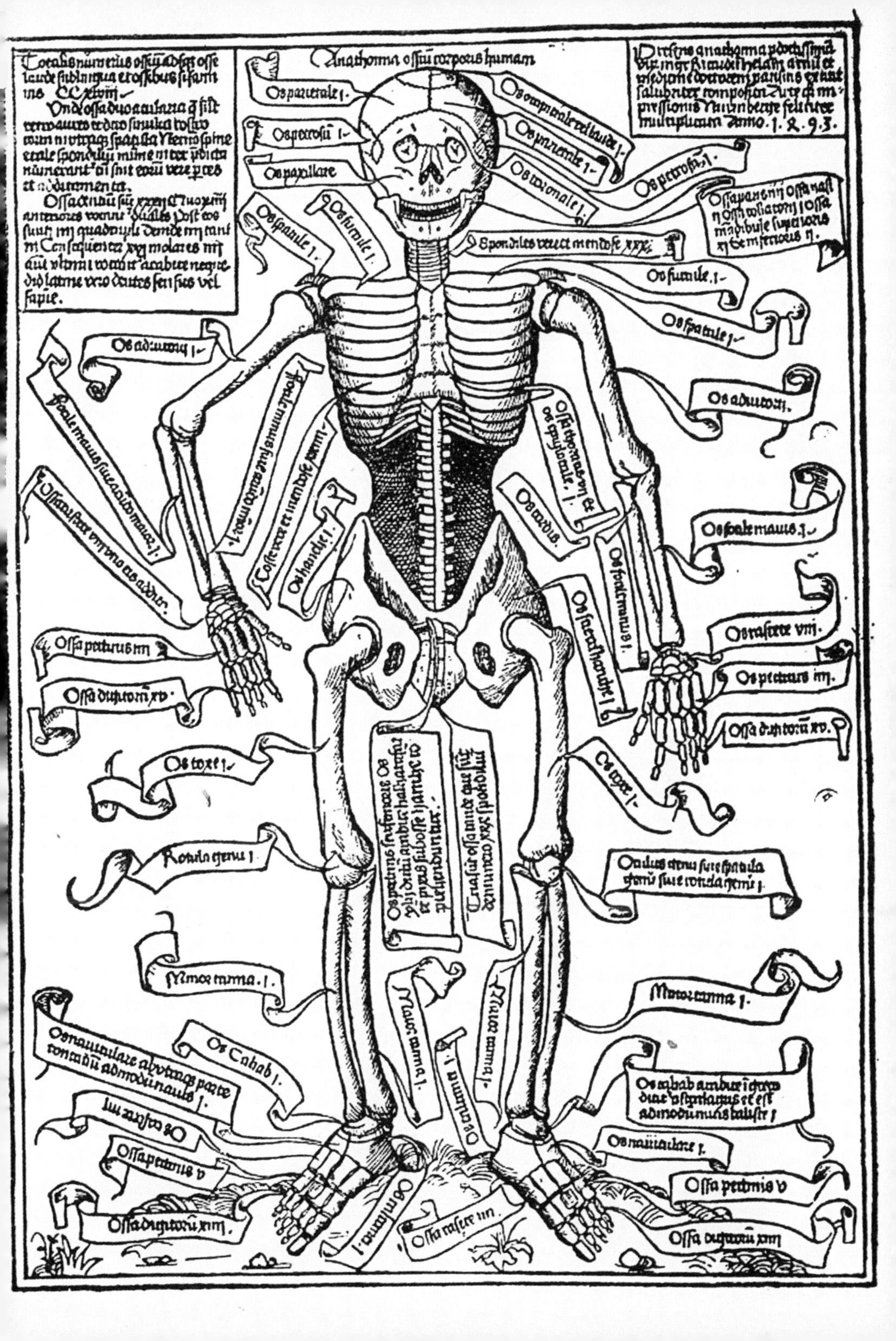
Anathomia ossium corporis humani
Os parietale 1.
Os petrosum 1.
Os coronale 1.
Os petrosum 1.
Os furcule 1.
Os spatule 1.
Os furcule 1.
Os spatule 1.
Os adiutorii 1.
Os adiutorii 1.
Os scale manus 1.
Os rasete viii
Ossa pectinis iiii
Ossa pectinis iiii
Ossa digitorum xv.
Ossa digitorum xv.
Os coxe 1.
Os coxe 1.
Rotula genu 1
Minor canna 1.
Minor canna 1.
Maior canna 1.
Maior canna 1.
Os Cahab 1.
Os nauiculare 1.
Ossa pectinis v
Ossa pectinis v
Ossa digitorum xiiii
Ossa digitorum xiiii
Ossa rasete iiii
Anno 1.4.9.3.

Liber de arte Distil landi de Compositis.

Das büch der waren kunst zü distillieren die

Composita vñ simplicia/ vnd dz Büch thesaurus pauperū/ Ein schatz d' armē ge/ närt Micariū/ die brösamlin gefallen vō dē büchern d' Artzny/ vnd durch Experimēt vō mir Iheronimo brūschwick vff geclubt vñ geoffenbart zü trost denē die es begerē.

around 1500 BCE. Hippocrates, whose word karkinos, meaning "crab", referred to a mass of diseased cells, thought cancer was caused by an imbalance of black bile. There was no cure, only palliative care, which took place in "temples of healing", an early form of hospice. These places used bathing with herbs, massage and music to soothe patients suffering from a range of complaints – from incurable diseases to inoperable battle wounds. Alas, the strongest painkillers they had were alcohol, willow and opium in its weakest form.

Perhaps the most elusive of all remedies is one for the good old common cold. Thousands of hours – and pounds – have been spent trying to find a cure, yet most of us still get the sniffles every year. Egyptian healers treated colds with magical spells, but other civilizations realized the condition was annoying but temporary, and merely strove to alleviate the symptoms. The ancient Greek remedy of alcohol, cinnamon (*Cinnamomum verum*) and honey isn't so very different from the soothing whisky, honey and lemon concoctions many swear by today. The Greeks called hyssop (*Hyssopus officinalis*) the "holy herb" because it ritually cleansed temples – and lepers. There was some doubt as to whether this was the same as the one we know as hyssop, but recent study shows the plant sometimes has a penicillin-producing mould on its leaves. Romans used it to relieve coughs. It was still being prescribed in 1655 by the mysterious W.M., who claimed to have seen service as chef to Queen Henrietta Maria before her husband, King Charles I, was beheaded and she was forced into exile. W.M. wrote a kiss-and-tell, *The Queen's Closet Opened*, published during the Commonwealth period, which claimed to reveal the secrets of the royal kitchen. His recipe To make Syrup of Hysop for Colds required a handful of hyssop, an ounce each of figs, "raysins", dates and French barley to be boiled from three pints of fair water down to a quart, clarified with two egg whites and boiled into a syrup with two pounds of fine sugar. The ingredients, certainly, would have needed royalty to afford them. Poorer people made do with hyssop tea.

Opposite An alchemist and his assistant use a double distillation still in an illustration from *Liber de arte distillandi de compositis*, Hieronymus Brunschwig, *c.* 1512. This manual describes practical techniques for manufacturing drugs and potions.

Galen, who liked using "opposites", treated colds with hot things such as fiery pepper, but other Roman physicians preferred the cool lettuce (*Lactuca serriola*). Culpeper agreed, suggesting strong infusions of wild lettuce (*Lactuca dregeana*).

The whole of the common eucalyptus (*Eucalyptus globulus*) or gum tree is useful to Aboriginal Australians. They use its bark for boats and the knobbly growths developed by mature specimens make rope-handled bowls. They treat cuts and wounds with the resin. Eucalyptus leaves are covered with oil glands that produce the plant's famous aroma, which acts as an insect repellent for the tree – and humans. Oil from the bark makes the tree highly flammable, which protects it because the fire burns quickly, then moves on, leaving the plant to regroup relatively undamaged. The antibacterial and analgesic properties of eucalyptus oil are reputed to treat everything from dry skin and fungal infections to muscle pain and toothache. Perhaps its most famous use, however, is for the common cold. While care should be taken not to consume it as it can be toxic, eucalyptus-infused steam is considered to alleviate congestion. We may not be able to cure the common cold, but after millennia of trying, we can at least make it more bearable.

Above Herbarium sheet of opium poppy (*Papaver somniferum*), collected at Kew, 2009. One of the few painkillers available in classical times.

Above Herbarium sheet of motherwort (*Leonurus cardiaca*), collected in Scotland, 1962. It was used for heart conditions, anxiety and menstrual cramps.

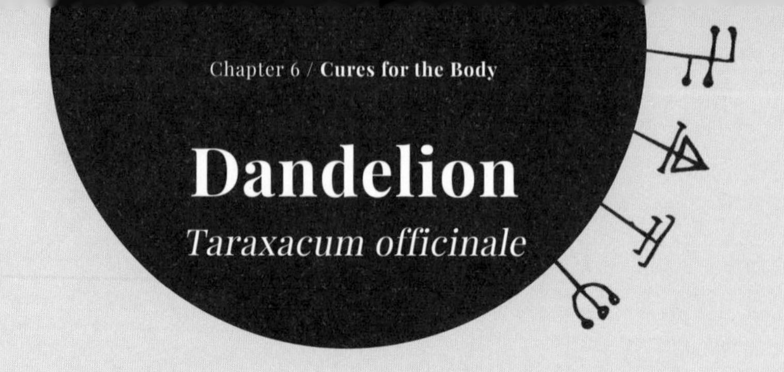

Dandelion

Taraxacum officinale

One of our most common wild flowers, the dandelion is recognizable even to people who claim not to know plant names.

What child hasn't tried to tell the time by blowing fluffy seed-head "clocks"? In some areas, the seeds were thought to be fairies and blowing on the clocks released the little folk from capture, so it was worth making a wish before making a puff. In other places, you made the wish if you caught the fairy, effectively holding it to ransom before letting it go again. Happily, another, more sinister legend – threatening children who blew away all the seeds with rejection by their mother – seems to have disappeared.

One of its many nicknames, "piss-a-bed" (*piss-en-lit* in French) provides a graphic playground explanation of what would happen to you if you picked one. This superstition does have a grain of truth; dandelion has been used as a diuretic for centuries, to cleanse the kidneys and open the urinary tract. Culpeper found the herb useful for a number of complaints, from rheumatism to weak hearts, and he also claimed it "helps one see farther without a pair of spectacles". He then goes on to say this information was better known abroad, implying that his enemies, the College of Physicians, kept knowledge about common plants secret so people would have to buy their expensive potions.

Other folk treatments used the milky latex from dandelion stems as a remedy for warts and sores – and the whole plant, when gathered on St John's Eve (23 June), was a potent witch repellent. In the meadow, eating dandelions was said to boost a cow's yield, even though cattle were not generally held to be fans of the flower's bitter taste.

A classic "pot herb", dandelion is still popular in dishes from Crete to Cairo. The roots are usually gathered in autumn and winter though for a hotter flavour, some harvest it in July. The young *dent-de-lion* ("lion's teeth") leaves are also used as a fiery addition to salads, something that would have been familiar to Tudor palates. During the Second World War, when coffee was scarce, people would roast and grind the roots as a substitute, a method popular with North American settlers in the nineteenth century. The UK's homemade wine craze of the 1970s rediscovered old recipes too, as millions of golden flower heads filled thousands of demijohns bubbling in garages up and down Britain, a process traditionally started on 23 April, St George's Day.

Opposite Dandelion (*Taraxacum officinale*) from *Flora von Deutschland Österreich und der Schweiz*, O.W. Thomé, 1885.

XIX, 1.
142. Compositae.
26. Lactuceae.
1
2
3
A
WM.
607. Taraxacum officinale Weber.
Gebräuchliche Kuhblume.

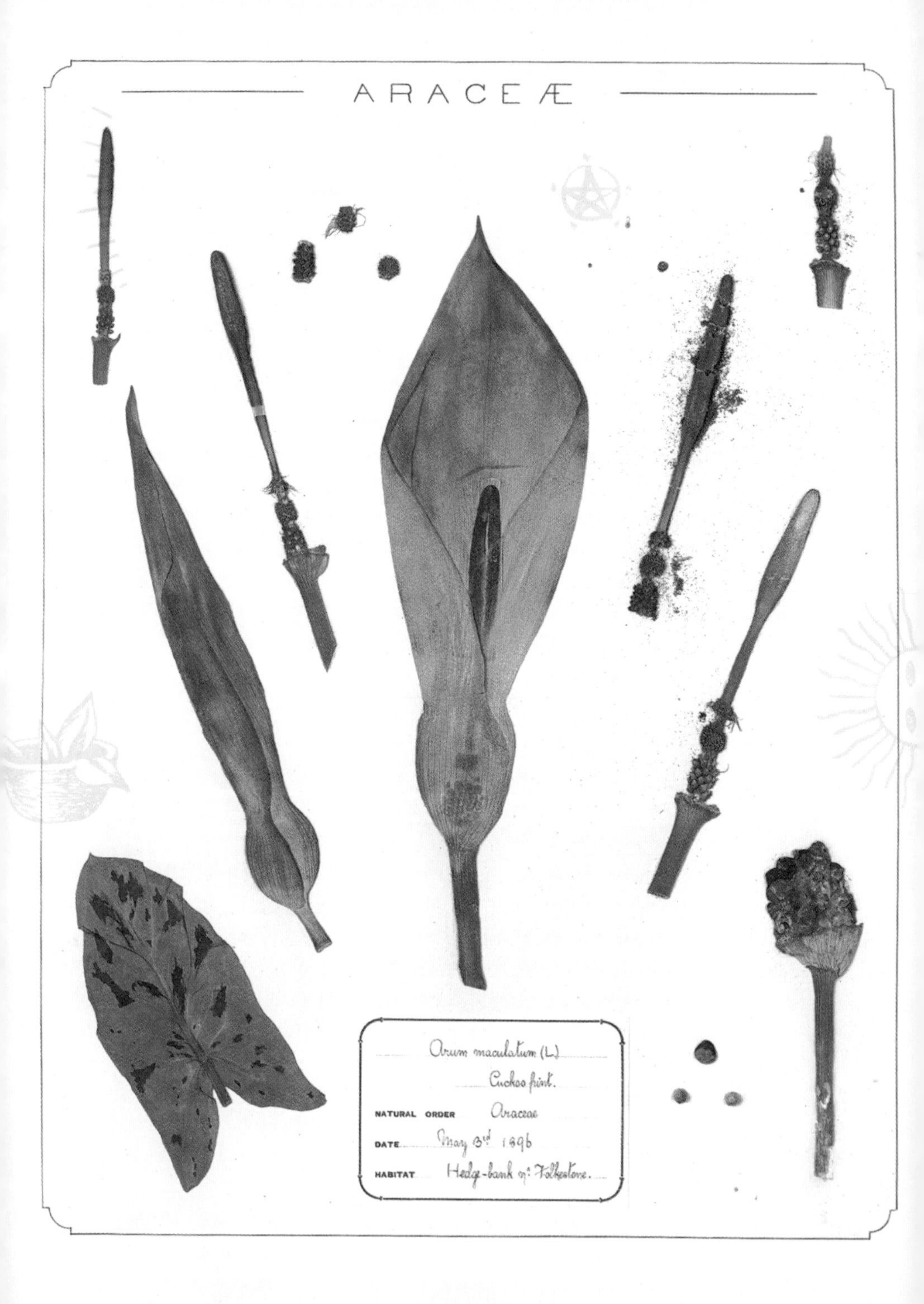
ARACEÆ
Arum maculatum (L)
Cuckoo pint.
NATURAL ORDER Araceae
DATE May 3rd 1896
HABITAT Hedge-bank nr. Folkestone.

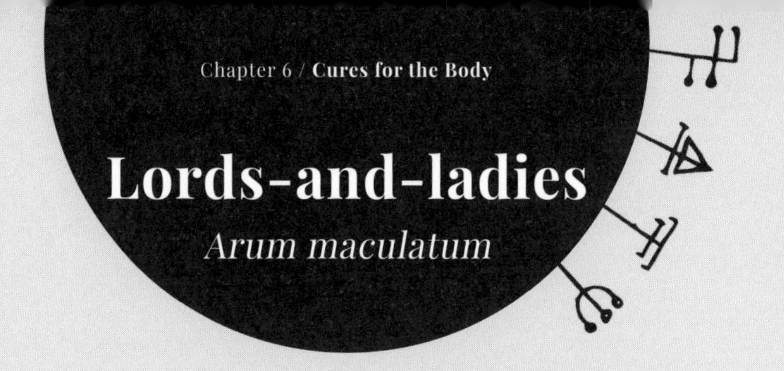

Lords-and-ladies

Arum maculatum

Some plants *look* poisonous for their own protection against predators. The eerie, slightly otherworldly lords-and-ladies, (*Arum maculatum*) is, however, as dangerous as it appears.

This striking, low-lying plant was said to have grown at the base of Christ's cross, so its dark green, arrow-shaped leaves are now permanently spattered purple from His blood. The creamy, yellow-green, hooded flowers (actually a bract, or "spathe") with their thick inner spike ("spadix") were sometimes called "parson in the pulpit" or, more worryingly, "Jack (or 'old man') in the pulpit", referring to Satan's country nicknames. The green and amber-red berries, standing on naked stalks like fairy traffic lights each autumn, lent the plant the name "snake's meat", because it was said vipers ate them to make venom.

Found in woods and hedgerows, lords-and-ladies has tiny, needle-like crystals in its cells that cause irritation to the skin. It also tastes horrid. This is a good warning sign, as it causes throat swelling, vomiting and breathing difficulties if ingested.

Given the plant's "masculine" properties, it's hardly surprising it also had sexual connotations. It was known as "wake-Robin" in some places, for its qualities as an aphrodisiac. If a man ate it, they would not sleep for love. If a girl even touched it, she invited (or risked) pregnancy.

It's hardly surprising the plant has been used as a poison down the ages, but it wasn't all bad. Also known as "cuckoo pint", the plant went by as many good aliases as ill (it has more than 150 folk names, including the vulgar but understandable "dog's dick"). It was also known in some places as "fairy lamps" because it was said to glow at night. The pollen does, indeed, emit a small amount of phosphorescence at dusk.

The root, properly prepared, was sometimes eaten and, in Victorian times, ground and marketed as "Portland sago", a substitute for arrowroot. Elizabethan ruff-makers stiffened lace with its starch; even after fancy collars went out of fashion, the ground root found use as an ingredient in "Cyprus powder" for styling eighteenth-century periwigs.

Lords-and-ladies held great interest for Nicholas Culpeper, who beat it into hot ox dung and applied it to gouty limbs. The bruised leaves, applied to boils and plague sores, drew out the poison within. A mixture of the root and leaf, boiled with wine and oil, relieved haemorrhoids. He also recommended boiling the juice of the berries in rose oil as a cure for earache. Few modern herbalists would agree.

Opposite Herbarium sheet of lords-and-ladies (*Arum maculatum*), collected in the UK by British Naturalist, James John Giles, 1896.

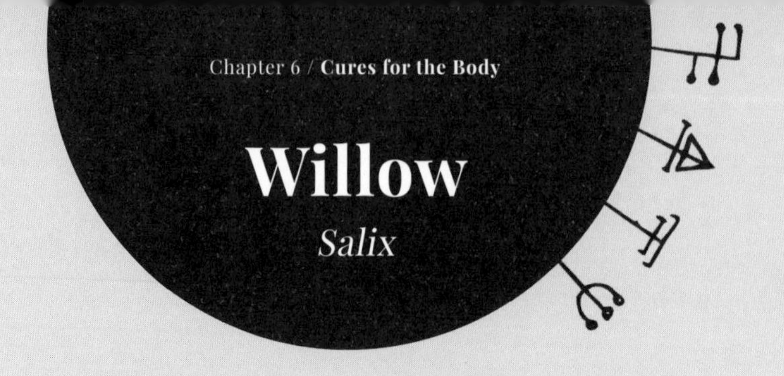

Willow

Salix

Willows have been associated with grief and loss since Orpheus carried willow branches with him on his way to the Greek Underworld.

In the Old Testament, Psalm 137 tells of the exiled Israelites hanging their harps on willow trees by the rivers of Babylon where they wept, giving weeping willow its Latin name, *Salix babylonica*. In early modern Britain, willow branches often decorated churches on Palm Sunday, genuine palms being in short supply. Willows are traditionally planted in cemeteries, too, across the world. Chinese people lay willows at the annual *Qingming* "tomb-sweeping" festival to honour the dead, as a symbol of regrowth and immortality. In Japan, the tree is associated with ghosts, though also with the beautiful, if heartbreakingly fragile, "flower and willow world" of the geisha. In Ireland, the "commoner of the wood" could, it was said, uproot itself and follow lonely travellers at night.

By the sixteenth and seventeenth centuries, willow leaves were associated with abandoned sweethearts and betrayal in love. You could unburden yourself to a willow, but should not expect the tree to keep its counsel; your secret would invariably be whispered to the four winds.

Traditionally a wood of the gallows, it was bad luck to burn willow, but a willow wand repelled evil or made a useful water-divination rod. It was said that English willow (*Salix alba*) made the finest cricket bats. For schoolchildren, it also made painful switches, known in Ireland as "sally rods".

The willow has been a symbol of misery for millennia, but the *Salix* genus deserves far better. It's not as though its properties as a nigh-on miracle drug weren't known about all the way back to the ancient Egyptians and Greeks. Hippocrates noticed the bark helped women in labour; Dioscorides suggested it for gout. Native American people made willow poultices for skin irritations and insect bites. Headache, toothache and earache were all said to be eased by chewing willow "withies" (twigs) and, in 1763, an Oxfordshire clergyman, Edward Stone, wrote to the Royal Society about how he had successfully treated a fever with willow bark. In the early nineteenth century, salicin was isolated from both willow bark and meadowsweet (*Filipendula ulmaria*) and it was converted into salicylic acid, the inspiration for a drug that relieves pain and can thin the blood. It was later found in much higher quantities in the sap and bark of willow, and a drug was created that could relieve pain and thin the blood. Although it is synthesized today, we all know it as aspirin.

Opposite Willow (*Salix alba*) from *Flora von Deutschland Österreich und der Schweiz*, O.W. Thomé, 1885.

168. Salix vitellina L. Dotterweide.

Umbelliferae.
1
2
3
4
5
6
7
8
9
10
11
12
A
WM.
Foeniculum capillaceum Gilib.

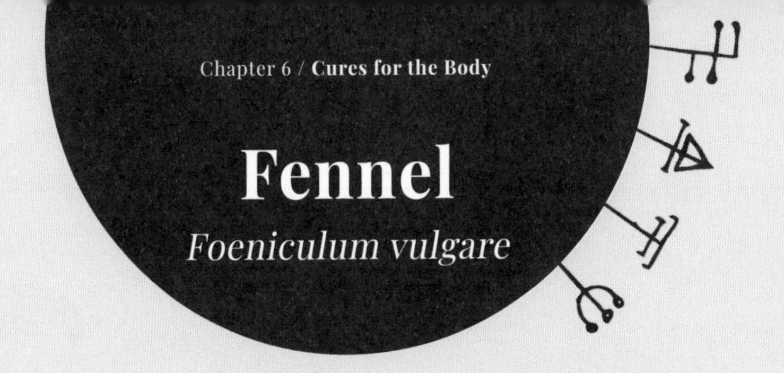

Fennel

Foeniculum vulgare

A Welsh saying, attributed to Iago ab Idwal, tenth-century King of Gwynedd, holds that "he who sees fennel and gathers it not, is not a man but a devil".

One of our most recognizable herbs, fennel is a favourite in cuisines from the Mediterranean to Asia, with its feathery fronds and distinctive aniseed flavour. A member of the Apiaceae family, along with carrots and celery, it originally grew in rocky places and cliffs, but it has naturalized in many places and, occasionally, is even considered a garden pest. Known as *marathos* in ancient Greece, fennel was sacred because Prometheus, having stolen fire from the gods, concealed it in the hollow stem of a fennel plant. The herb was used in rituals, including sprouting it in pots during the annual eight-day Adonia, the festival of Adonis. The town Marathon was named for the plant, as it grew so heavily on the local slopes.

Fennel was one of the nine sacred herbs of the Saxons and later thought to be consumed by snakes before shedding their skins, making it a symbol of renewal. Pliny the Elder discusses more than 20 ailments that could be cured with the herb and the Emperor Charlemagne commanded it be grown in all his imperial and monastic gardens. The English King Edward I bought eight-and-a-half pounds of it in a single month. A favourite Tudor recipe for "comfits" involved the painstaking sugar-coating of fennel (and other) seeds to be nibbled as digestifs after a meal. *The Good Huswifes Jewell*, from 1585, recommends a drink made from fennel "to make one slender".

Chewing the seeds relieved indigestion and bloating, eased flatulence and stopped hiccups. Culpeper recommended it to increase the milk of nursing mothers, "provoke urine" and "break a stone", while the seed boiled in wine could ease the stomach of someone who had eaten poisonous herbs or mushrooms. The whole herb, distilled, was considered good for the eyesight, but it was useful in so many other ways. Hung over doorways at midsummer, it would ward off witches; attached to horse bridles, it deterred flies; strewn on floors, it was a flea repellent; and plugging keyholes with fennel kept out ghosts.

For all its many properties, some people preferred to forage fennel in the wild rather than risk bad luck in growing it; an old adage warned, "sow fennel, sow trouble". Perhaps they were all too familiar with the dozens of seedlings that crop up each spring in a garden that cultivates the herb.

Opposite Illustration of fennel (*Foeniculum vulgare*) from Köhler's *Medizinal Pflanzen*, 1897.

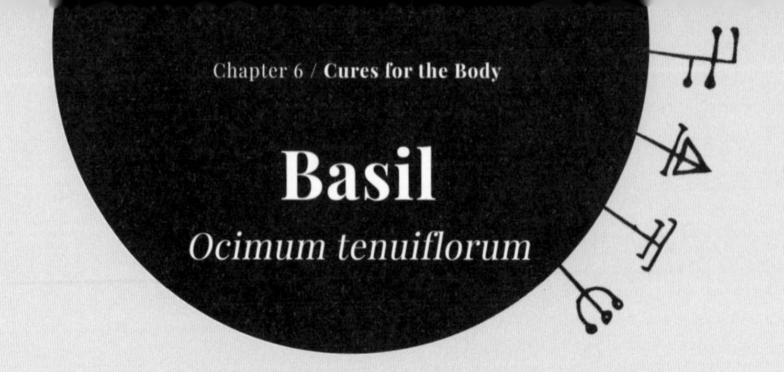

Basil

Ocimum tenuiflorum

For a herb so beloved today, sweet basil started out poorly. The ancient Greeks were unimpressed with it as a medicinal herb, though it was said to have many properties they valued.

Dioscorides claimed that, taken in large quantities, it would cause dim-sightedness, soften the bowel and set gases in motion. It was a diuretic, it stimulated lactation and it was generally hard to digest. Associated with death in Crete, poor basil belonged to the Evil One. In other parts of Europe, it was considered a witches' herb. If you left a basil leaf under a pot, it would turn into a scorpion.

Pliny and Arabian physicians came out in defence of the plant but the jury was still out by the time Culpeper was writing: "This is the herb which all authors are together by the ears about and rail at one another (like lawyers)". According to the Greeks, to flourish, basil needed to be verbally abused while it was sown. It has been associated with discord ever since. In French, to *semer le basilic* ("sow basil") still means to rant and rave.

For all this, however, basil was an extremely useful herb, for flavour, strewing and as a powerful bug repellent, warding off flies, mosquitoes, cockroaches and other insects. Many people in the Mediterranean still grow a pot of basil by an open window for the same purpose.

The tide began to turn with stories of St Helena's discovery of the True Cross. Helena lived in the fourth century, but tales of her miraculous pilgrimage to Jerusalem grew over several centuries. It was said that God had promised her a sign to lead her to Christ's cross. After searching for many days and nights, she saw a sweet-smelling plant growing on an otherwise bare hill. Underneath it she found the holy relic. Even today, the Greek Orthodox Church often uses basil in holy water.

Slowly, sweet basil became associated with more pleasant superstitions. In Moldova, a lad that accepted a sprig of basil from a girl would love her forever. Mexican people carried a few leaves in their pockets to attract luck and money, and in Italy, basil finally became the herb of love. *Tulasi* ("holy basil", *Ocimum tenuiflorum*) grows in Hindu temples and homes as a sacred aromatic perennial. A manifestation of the goddess Lakshmi, it must never be picked.

In some places, it is said that if a woman puts a pot of basil outside her room, it's a sign that she is ready for love. Of course, she might just be trying to keep the insects at bay.

Opposite Basil (*Ocimum basilicum*) from *La botanique mise à la portée de tout le monde*, 1774.

Le Basilic

Ocymum Basilicum. L. S. P.

de Nangis Regnault f. Ital. Basilico. Angl. Basil. Allem. Citronen Basilien.

a. *Allium sativum.* Ail aulx. Knoblauch.
b. *Allium campestre.*
c. *Allium juncifolium luteum.*

Garlic

Allium sativum

Garlic is one of the world's most popular foods, spices and medicinal plants, and has been since it made its way from Asia to the West.

It was revered by the ancient Egyptians, who gave it to labourers building the Great Pyramid of Giza to give them strength and ward off illness. Ancient Korean people ate it before going up mountains, as tigers were thought to shy from the smell.

Equally averse to the plant were evil spirits. People from many cultures would eat it before travelling at night, and Ancient Greek midwives hung bulbs in the birthing chamber to keep evil at bay. The tradition linking garlic with vampires began in Romania, where the bulb was originally a general catch-all against demons, witches and sorcery, but slowly specialized. Garlic cloves were stuffed into the orifices of corpses suspected to be vampires, and in China and Malaysia, the juice was smeared on children's faces to prevent attack while they slept.

A member of the Amaryllidaceae family, garlic is full of vitamins and minerals, but it is thought the bulb's main therapeutic power comes from the active ingredient allicin. When garlic is crushed, sulfur-containing constituents such as alliin come into contact with the enzyme alliinase and are converted into different chemicals, including allicin. Allicin is present in other alliums, such as onions and leeks, but in far smaller quantities. Anti-inflammatory and antiseptic, garlic has long been an item one should never go a day without.

The ancients didn't know about allicin, but they did know whatever was in the plant worked. Hippocrates recommends garlic for infections, wounds, cancer and leprosy. Dioscorides liked it for heart problems, as do many today. Pliny the Elder lists no fewer than 61 remedies involving the bulb.

Folklore didn't always need garlic to be taken in a regular fashion. Wearing garlic inside one's shoes prevented whooping cough; burying a bulb in the garden was a sure-fire cure for measles. Garlic also formed part of the notorious "vinegar of the four thieves" during plague time (see page 170).

Worms, cuts, phlegm, ulcers, spots, the bite of mad dogs – garlic was, according to Culpeper, "the poor man's treacle, being a remedy for all diseases and hurts". He did caution moderation, however, warning that garlic's heat was "vehement" and that while it would help those "oppressed by melancholy", "in choleric men it will add fuel to the fire". Garlic was also used as an antiseptic by all sides in the Second World War, and is still used today as a flavoursome route to a healthy life.

Opposite Garlic (*Allium sativum*) from *Phytanthoza iconographia*, J.W. Weinmann, 1737.

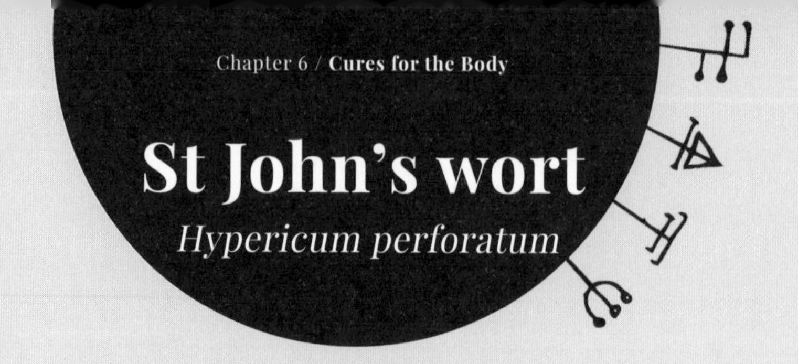

St John's wort

Hypericum perforatum

The sunshine-yellow flowers of St John's wort are such a common sight in many a garden and even car park that we sometimes forget to look at it the way our ancestors once did.

For centuries, *Hypericum* has been associated with St John the Baptist, whose blood was reputed to flow through the plant as a red-pigmented oil. It was said the oil appeared as bloody spatters on the leaves on 29 August each year, the anniversary of the saint's beheading. Other legends claim the herb was used by the knights of St John of Jerusalem to heal battle wounds during the Crusades.

St John's wort was used as a painkiller and to ward off evil spirits: "insane" people were given drinks containing the oil to calm them down. The herb was particularly revered in Scotland. In some areas, it was treated with holy smoke then hung in the home, cowshed and dairy. In Aberdeenshire, sleeping with a piece of St John's wort under your pillow would bring blessings from the saint himself, along with pleasant dreams. People in the Hebrides took this one step further, wearing the herb in their underclothes to ward off "second sight", enchantment and death, and to bring peace and plenty. It only worked, however, if the plant had been found by accident. Searching for it weakened its powers.

The "fairy herb" was best gathered on the morning of St John's Day, 24 June, with dew still upon it. If a girl picked a sprig and it was still fresh the next morning, so were her chances of finding a husband. A woman walking naked to harvest the herb would conceive within the year. In other places, the plant was hung around homes on St John's Eve, to keep away ghosts, demons and thunderbolts. In Wales, a sprig was gathered for each member of the family, named and hung up: the first to wither heralded that person first to die. On the Isle of Wight, people had to be careful not to step on the shrub in case a fairy horse kidnapped them, made them ride to exhaustion, then abandoned them far from home.

Although St John's wort has been used as a diuretic, a treatment for worms, nerve irritation, coughs and bruises, and as an ointment to close wounds, it is most famous for its use with mood disorders. Culpeper commended the herb against melancholy and madness; it is still used today for its antidepressant properties and to combat seasonal affective disorder.

Opposite St John's wort (*Hypericum perforatum*) from *Bilder ur Nordens Flora*, *c.* 1901–1905.

MANSBLOD, A. HYPERICUM PERFORATUM L.
B. HYPERICUM MACULATUM CR.

PLANTS OF UGANDA, AFRICA
Herbarium of the Arnold Arboretum, Harvard University
No. 446 Mrs. M. V. Loveridge Jan. 31, 1939
Nymphaea caerulea Savigny s.l.
Lake Mutanda, above Mushongero; alt. 5925 ft.
In lake. Flower purple with yellow centre.
Flora of Tropical East Africa
Det. Date 29.1.88

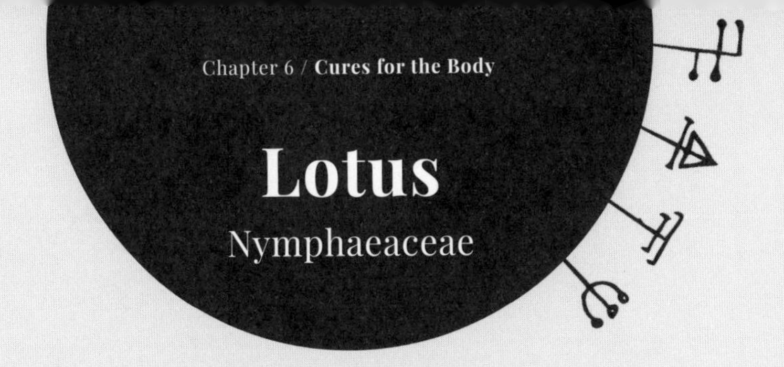

Lotus

Nymphaeaceae

There are few plants held as widely sacred as the lotus is – a mythical plant, associated with creation myths, death and rebirth.

Whether the native European yellow water lily (*Nuphar lutea*) or the blue lotus (*Nymphaea nouchali* var. *caerulea*) and white lotus (*Nymphaea lotus*) of eastern traditions, it is venerated around the world. The ancient Egyptians called the blue lotus the "flower of the sun" because of its yellow "eye" and habit of rising up from the water at dawn. The god Horus was reborn each day from the lotus after having spent the night in the closed flower. The first lotus rose from chaos at the beginning of time, revealing the creator in the form of a scarab beetle.

The lotus appears in countless tomb paintings; its hieroglyph also means the number 1,000. The white lotus, representing the resurrection of the god Isis, was the symbol of Upper Egypt. Mildly narcotic, both varieties were used in medicine, but not always for beneficial purposes. One unguent, made with lotus leaves, was to be applied "to the head of a hated woman", to make her hair fall out.

In Hindu tradition, Vishnu made a lotus with 1,000 golden petals, upon which Brahma the creator sits. The lotus also represents the womb of creation, from which the goddess Padma was born. It is the emblem of paradise in Japan. In Buddhism, the lotus is a powerful image of enlightenment, symbolizing the control of the spirit over material desire. It is usually depicted partially open, suggesting that full understanding is not yet achieved.

A symbol of beauty to the Greeks, the yellow water lily was used to produce opium-like effects. More prosaically, it treated vaginal discharge and, in a reversal of the ancient Egyptian curse, water lilies with white flowers were used to cure dandruff.

Dioscorides suggests using the root of the water lily to produce impotence and cure erotic dreams; John Gerard has a similar idea, recommending it "against fleshly desire" and to cure "the overflowing of seed which cometh away by dreams". It was also good for the "bloody flux" (dysentery).

In Greek mythology, the lotus-eaters were a race of islanders dedicated to luxury and idleness. Permanently soporific thanks to their consumption of a strange plant, they encouraged Odysseus's men to indulge too, costing the crew valuable time. No one knows which plant Homer was referring to, but he may have based his lotus on jujube (*Ziziphus lotus*), a variety of buckthorn.

Opposite Herbarium sheet of a purple lotus (*Nymphaea nouchali* var. *mutandaensis*), collected from Lake Mutanda, Uganda, 1939.

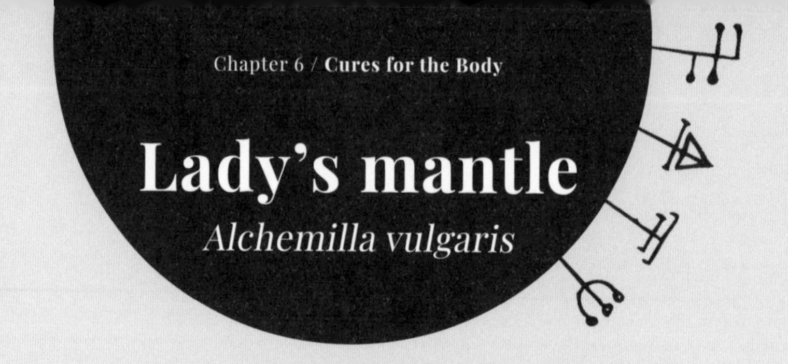

Lady's mantle

Alchemilla vulgaris

Gardeners tend to dismiss *Alchemilla* a little these days, considering it pretty enough but, in its cultivated form (*Alchemilla mollis*), mainly useful as ground cover.

In the past, however, "Our Lady's mantle" was much revered. Dedicated to the Virgin Mary, the plant has leaves that form a frilly, fan-pleated and scalloped "cloak", fit for the mother of God. In the mornings, they are studded with diamond-like droplets of dew, giving the plant yet another nickname, "dew cup". These crystal-clear, sparkling results of the plant's "superhydrophobic" (highly water-repellent) surface were considered holy – water that could cure all illnesses. It was particularly prized by alchemists, who used it in their magical potions, believing the magical dew would help them create the mythical Philosopher's Stone. The Arab word *alkemelych* ("alchemy") gives the plant its Latin name.

Lady's mantle is a perennial herb, native to Europe, Northwest Asia, the eastern United States and even Greenland. A low-lying plant with a fizz of acid-green flowers, it was also known as "lion's foot" or "bear's foot" because of its club-forming, spreading roots. It grows in pastures, at the edges of woods and in hedgerows, and was popular with dyers for its subtle, yellow-green hue. It has always been considered a "woman's herb", ruled by a number of female goddesses, from Freya/Frigg, Norse wife of Odin, to the Christian Mary. Culpeper associates the plant with Venus, Roman goddess of love and beauty, and it was often used in cosmetic preparations. Dew collected from the centre was supposedly a powerful beauty lotion, while a pillow filled with the herb encouraged beauty sleep.

An anti-inflammatory water distilled from the highly astringent leaves was said to reduce pores, dry out acne, lighten freckles, relieve inflamed eyes, stop bleeding, help vomiting and flux and ease bruises. Taken orally for 20 days, it could help conception, regulate the menstrual cycle and ease the symptoms of menopause. It is still prescribed by some herbalists for excessive menstrual bleeding. If a woman was suffering "over-flagging breasts", Culpeper recommended she drink the water, while simultaneously applying it topically, to reduce and harden them. It was also whispered the plant's astringency would reduce "slipperiness of the womb" and reverse sterility. In other, darker places, a concoction of lady's mantle was said to contract the genitalia in women who wished to appear virgins on their wedding nights.

Opposite Lady's mantle (*Alchemilla vulgaris*), from *Flora von Deutschland Österreich und der Schweiz*, 1885.

IV,1. 105. Rosaceae.
5. Poterieae.
1
2
3
4
A
B
Frauenmantel.
410. Alchemilla vulgaris L.

Love potions

What are herbs for if not for finding one's true love and then snaring them? Throughout this book herbs are used (chiefly by young girls) to dream of their future husbands.

Love spells and potions fill many a modern witch's herbal, but historical recipes for love potions were often complex, hard to make and, frankly, repulsive. The thirteenth-century bishop Albertus Magnus penned many books, on subjects from logic to alchemy, but it's unlikely he wrote the sixteenth-century *The Boke of Secretes of Albertus Magnus*. This work suggested periwinkle (*Vinca*), "houselyke" (houseleek, *Sempervivum*) and "worms of the earth", ground to a powder and added to meals produced love between a man and his wife. One Persian love potion combined pigeon broth with cloves (*Syzygium aromaticum*), laurel seeds (*Laurus nobilis*), thistle (*Cirsium*) and sparrow-wort (*Thymelaea hirsuta*).

In the Bible, the *Song of Solomon* makes several references to mandrakes (*Mandragora*) as aids to love. With this in mind, medieval country fairs were only too happy to supply "mandrakes" to love-lorn locals. Alas, they often turned out to be carved turnips.

William Shakespeare knew when the folklore of his home county, Warwickshire, could shift theatre tickets. In *A Midsummer Night's Dream*, the fairy king Oberon instructs Puck to fetch a herb that will make Queen Titania fall in love with the first creature she sees. "Love-in-idleness" was a West-Midlands name for heart's ease (*Viola tricolor*), once white but now "purple with love's wound". Shakespeare liked the plant so much he even named a heroine for it. The unromantic Culpeper was less moved by *Viola*'s charms, describing the plant as cold, viscous and slimy, though worth keeping in mind to treat venereal disease.

Among other names, southernwood (*Artemisia abrotanum*) is also known as "lad's love", "kiss-me-quick", "nobby-old-man" and, slightly alarmingly, given the previous titles, "maiden's ruin". A young man with love on his mind would wear a sprig in his buttonhole, very obviously "sniffing it" as girls walked by. If any of them noticed this strange behaviour and didn't run in the opposite direction as fast as possible, he could present her with the herb. To the giggles of the rest of their friends, the two would take their first courting-stroll.

The Victorians were famous for their complicated *Language of Flowers*, where a secret message could be spelled out using the different blooms in a posy. Who knows how many people actually bothered to stand in a florist shop with a floral dictionary, but sending flowers of any kind is still a pretty good way to most people's hearts.

Pro Amore [for love]

Take Valerian an herbe, and put in a glass of Bier or wine and give the same to whom thou wish love thee extremely.[3]

Lavandula latifolia.

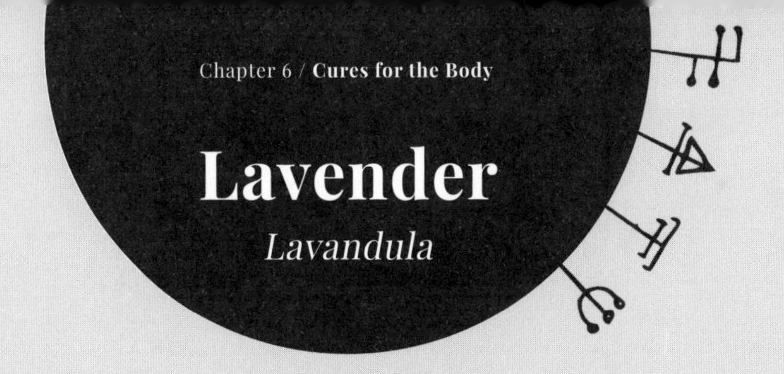

Lavender

Lavandula

Lavender is one of our best-beloved herbs, prized mainly for its aromatic and cleaning properties since antiquity – its name comes from the Roman *lavare* ("to wash").

The Egyptians used it in wax cones that melted slowly over their heads, releasing the perfume, or as a fragrance for mummy shrouds. One Christian legend (shared with rosemary) has the Virgin Mary washing the Holy Family's clothes and laying them on lavender bushes to dry. The flowers were stained blue and gained a heavenly fragrance from the experience. We still use lavender in everything from deodorant to floor wax.

A woody, shrubby plant with narrow, silver leaves and purple-blue, spiked flowers, lavender was believed to be powerfully repellent to tigers and lions. Less ambitiously, we use it to deter moths. When humans inhaled the scent, they could see ghosts, but it was more often worn by children as a talisman against the Evil Eye. During the Middle Ages, it became a herb of romance. Girls snuck lavender under the pillows of men they liked to turn their thoughts to romance. Charles VI of France clearly liked the idea; in 1387, he had cushions stuffed with the herb.

Lavender is used in the kitchen, but relatively rarely, as the taste is strong and highly perfumed. Sugar is sometimes flavoured with it – one dessert, now sadly lost to time, consisted of fruit and bonbons threaded onto lavender stalks.

As a herbal medicine, however, lavender came into its own. It had seemingly opposing qualities by being both stimulating and sleep-inducing. Also known as "Sticadove", it appears in all the great herbals, used in posies to ward off plague and given in bundles for women in labour to squeeze for strength. In William Turner's *A New Herball* (1551), he recommends dried lavender flowers be quilted into a cap. If worn daily, it was good for head colds and comforted the brain.

Lavender could also be administered as an extracted oil. Culpeper warns that it is "of a fierce and piercing quality and ought to be carefully used". Just a few drops relieved complaints such as stomach problems, passions of the heart, fainting and blocked liver or spleen.

In 1910, French chemist René-Maurice Gattefossé burned his hand and, looking for something to put on it, alighted on some lavender oil. It was said that the wound healed without scarring and he went on to use it for soldiers to aid healing in the First World War. Gattefossé later delved deeper into the concept of healing aromatic oils, and aromatherapy was born.

Opposite Lavender (*Lavandula latifolia*) by German botanist Friedrich Gottlob Hayne, 1822.

Apple

Malus

Entire books have been written about the folklore of the apple, whether the wild "crab" apples, sweet "eaters", sharp "cookers" or tart cider apples.

Described by Theophrastus as one of the most civilized trees, apples have been so successfully exported by humans that they are common throughout the world.

Of course, there are plenty of stories. In Greek mythology, Dionysus created the apple to present to Aphrodite, the goddess of fertility; she famously gave it away to Paris of Troy. Norse myth tells how the trickster Loki stole the apples that Ithun, the goddess of youth, had given the gods to inhibit old age. The apple is still associated with immortality.

Apples, especially in cider counties, were the life source of the countryside, and the health of a family was considered directly related to that of the tree. If the sun shone through the branches of apple trees on Christmas Day, there would be an abundant crop. This outcome could be further enhanced by "wassailing", or drinking the health of the trees, which usually took place on the twelfth day of Christmas. Also known as "apple howling" in some districts, the idea was to make as much noise as possible, to wake the trees from slumber. People would shoot muskets into the branches, shout, blow horns, drink ale and eat cake.

Apples were "christened" on St Swithin's Day, 15 July. If you tried to eat the fruit before that date you would be very ill. An apple flowering out of season foretold death, as did a single fruit left on the tree after the first frost, but some lovers carved their initials into the young apple, believing their affection would grow with the fruit. Orchards were protected by "Lazy Laurence" or "Awd Goggie", the spirit living in the oldest tree. In the English county of Somerset he was known as The Apple Tree Man.

Halloween was a time for apple divination. Peeled in a single strip and thrown over the shoulder, the skin would form the initial of a future husband. You might also twist an apple stalk while reciting the letters of the alphabet; the letter uttered at the break would be "the" initial. Apple pips thrown into the fire foretold whether your love would be true. If they popped and fizzed, so would the relationship.

Apples are the traditional fruit of health. They were said to cure warts, clean teeth, cure constipation and beautify the face. Rotten apples were applied to chilblains, while an apple placed in the same room as someone with smallpox "caught" the disease in their place. As it withered, so did the patient's symptoms. Apple cider vinegar is still served with honey and lemon for colds.

Opposite Hawthornden apple by Johann Hermann Knoop, from *Pomologie, ou description des meilleurs sortes de pommes et de poires*, 1758.

Galo-Bayeux.

De l'Imprimerie de Langlois.

The Wild Flora of Kew Gardens

Name: *Digitalis purpurea* L.

Vern. name: Foxglove

Location: West Arboretum: amongst shrubs in the southern part of the Rhododendron Dell (zone 228)

Notes:

Foxglove

Digitalis purpurea

"Goblins' thimbles", "fairy weed", "snoxums", "snompers", "fairy's petticoat" – the local nicknames for *Digitalis* have to be some of the most evocative of all plants.

As filtered sunbeams pick out the swathes of foxglove purple in woodland glades, the world heaves a sigh of relief: summer is almost here. It's a clever plant; the female flowers at the bottom of the stems contain the most nectar, persuading the bees to visit there first and then work their way up the spike to the male flowers, pollinating as they go. In olden times, the fat flower buds did have to survive being "popped" by children (giving them the name "pop docks"); it's less of a problem now, as is the fear of "hearing the flower bells ring", a warning of imminent death. In Scotland, foxgloves were even called "dead men's bells". Perhaps the foxes still listen out; it is said the bells ring to alert the arrival of huntsmen. Medieval fables tell of a fox called Reynard, to whom the fairies give foxglove flowers to silence his paws as he raids chicken coops. Perhaps Beatrix Potter knew foxgloves are associated with insincerity when her foxy-whiskered gentleman called on Jemima Puddleduck.

It was unlucky to bring foxgloves, especially white foxgloves, into the house, as it encouraged witches. They could be useful, however, in a somewhat risky method of identifying a changeling. The child was given three drops of foxglove juice, put on a shovel and swung out of the front door three times, the parents crying, "If you are a fairy, away with you!" If the child was a changeling, it would die. If it was human, it would be traumatized for the rest of its life. The child would be ill, at the very least. *Digitalis* contains toxins, including cardiac glycosides, which increase heart rate. Nausea, headaches, diarrhoea and visual, heart and kidney problems are just some of the symptoms caused by ingesting the plant. Nevertheless, the leaves were useful to bind around fresh wounds. Placed in a child's shoes, they were said to guard against scarlet fever.

While those cardiac glycosides could be fatal, others have been developed into pharmaceutical drugs. It's possible the Egyptians knew about foxglove's ability to stimulate the heart – but in 1775, Dr William Withering, searching for treatments for dropsy (oedema), began systematic trials using *Digitalis*. The resulting *An Account of the Foxglove and some of its Medical Uses* (1785) proved a game changer in the treatment of certain heart conditions. His memorial, in St Bartholomew's churchyard in Edgbaston, is carved with foxgloves.

Opposite Herbarium sheet of foxglove (*Digitalis purpurea*) collected at Kew, 2009.

Chapter 7:
Plants and the Heavens

Ancient astrology emphasized the supernatural influences of the heavens, ascribing gods to the stars and planets. The Egyptians and Babylonians divided the heavens into 12 sections; six northern signs and six southern, very similar to the zodiac we know today. The stars ruled everything, from monarchs to marigolds.

The seven planets, known to the Romans as Jupiter, Venus, Saturn, Mars and Mercury, along with the Sun (Sol) and Moon (Luna), were said to be ruled by the gods that bore their names. Through these planets, the gods ruled the sky and earth.

Nations and wars were decided by the heavens, so it wasn't long before people started believing the fates of individuals were also governed by the stars. If humans were influenced by the zodiac, though, wouldn't all things – animal, vegetable or mineral – fall under the same rules? After all, noted Pliny, other civilizations were governed by the skies above. He used the example of rituals made by British Druids, gathering mistletoe when the Moon was in its sixth day. Astronomy and astrology (not then considered separate disciplines) were widely practised in Arabian countries, too, to an even greater depth, and many of their studies were translated into Latin for Western use.

By the end of the sixteenth century, laws in some European countries required physicians to calculate the position of the Moon before committing a patient to surgery.

In the seventeenth century, astrology took on a more "scientific" nature, using "logical" reasons why the heavens had an effect on the body. Astronomers observed meteorological disturbances through their telescopes and figured these must impact the plants and creatures of Earth. The disturbances could be big, such as eclipses or transits of one planet across another, or so small that no one could see them, but they all had an effect. The physician Richard Mead (1673–1754) believed the Moon's gravity affected bodily fluids the same way it does the tides of rivers and oceans. This had, he pointed out, potentially catastrophic effects, from inducing fevers to causing hysteria. It even explained the ancient idea of madness or "lunacy".

Botanical astronomy was the belief that each plant grew under the influence of a particular heavenly body. The fifteenth-century Swiss writer Paracelsus (approximately 1493–1541) went a step further, considering each plant a terrestrial star reflected in the heavens. Much medieval folk medicine dictated when a herb could be collected for maximum potency depending on when the plant's reigning planet was visible in the sky.

Individual body parts were also influenced by the stars – for example, Venus ruled the kidneys and digestive system, while Saturn was governor of the bones, veins and skeleton. Ovaries were, of course, the domain of the Moon.

Nicholas Culpeper subscribed to the philosophy that believed each disease was caused by the movements of a planet, a concept that dated back to the ancient Greeks. There were two ways of curing the problem: either by using "sympathetic" herbs, "governed" by the same planet as the disease – or plants ruled by planets directly opposite the disease on the zodiac chart. Of course, it was much

Opposite Sunflower (*Helianthus*), 1867. The Latin name refers to Helios, the Titan god of the Sun.

more complicated than just choosing the "correct" herb. The exact date and time of the onset of the condition was important too. Practitioners needed to create an astrological chart or "decumbiture" (from the Latin *decumbo*, meaning "to lie or fall down") for the illness itself, from the precise moment the person felt poorly, before they could be certain which herbs would do the trick.

In order for his readers to diagnose conditions themselves, Culpeper wrote a companion volume to his bestseller *The English Physician*, entitled *Astrological Judgement of Diseases from the Decumbiture of the Sick*. Alas, it was published after his own death at the tragically young age of 37. His ideas weren't universally accepted; his old adversaries, the College of Physicians, were particularly critical, and they held sway among a large proportion of regular people for many years.

The Greeks and Romans, who sacrificed to both Sol, the god of the day, and Luna, the goddess of the night, believed that the sap in plants waxed and waned with the Moon. The jury is still out, but experimentation aboard the International Space Station suggests that the Moon's gravity might affect plants after all. There is little evidence that planting by the phases of the Moon has much effect, but many scientists are now keeping an open mind.

The Sun was hot and masculine, his flowers gaudy, bright and golden: dandelion (*Taraxacum*), marigold (*Calendula officinalis*) and sunflower (*Helianthus*). Herbs of the Sun were associated with strength, the heart and the body's vital fluids.

Before the summer solstice, the Sun was said to be ascendant and plants grew with vigour. After midsummer, the rays softened and plants began to wane as the Moon grew in influence. This was reflected in a smaller, daily rhythm: sap rose in the mornings, peaked at midday, and then fell away in the afternoon before sinking again at night.

Known to generations of gardeners as the "Parish Lantern", moonlight was often the only time of the day manual labourers could work on their own plots. The lunar month is roughly 29 days, and each waxing and waning needed to be used for different jobs. An April Moon traditionally saw the germination of seeds, sown into soil ploughed during the time of a January Moon, when the frost had already broken down the hard soil. The May Moon was good for planting out, while the Harvest Moon (the nearest full moon to the autumnal equinox) was so strong, workers could continue to bring in the crop by night. The December (or Christmas) Moon was, traditionally, the time to prune fruit trees.

The Moon was associated with cold, as her silvery light was best seen in a clear sky, bringing similarly shimmering frost. Her flowers were feminine, influencing the menstrual cycle and childbirth. Any plants that bloomed by night were considered particularly magical, especially if they were white-flowered, like the heavily-scented night jasmine (*Cestrum nocturnum*) or had silver leaves, like the mysterious mugwort (*Artemisia*).

The exact timings of the Moon's phases could be calculated from almanacs. One of the most famous was first published in 1697 by self-taught physician Francis Moore, astrologer at the court of King Charles II. To start with, Moore just included weather forecasts but, three years later, he published *Vox Stellarum* or "*The Voice of the Stars*" with astrological predictions, which proved much more popular.

Opposite Volvelle or movable zodiac chart from fifteenth-century manuscript *The Guild-book of the barber-surgeons of York*, *c*. 1475–99.

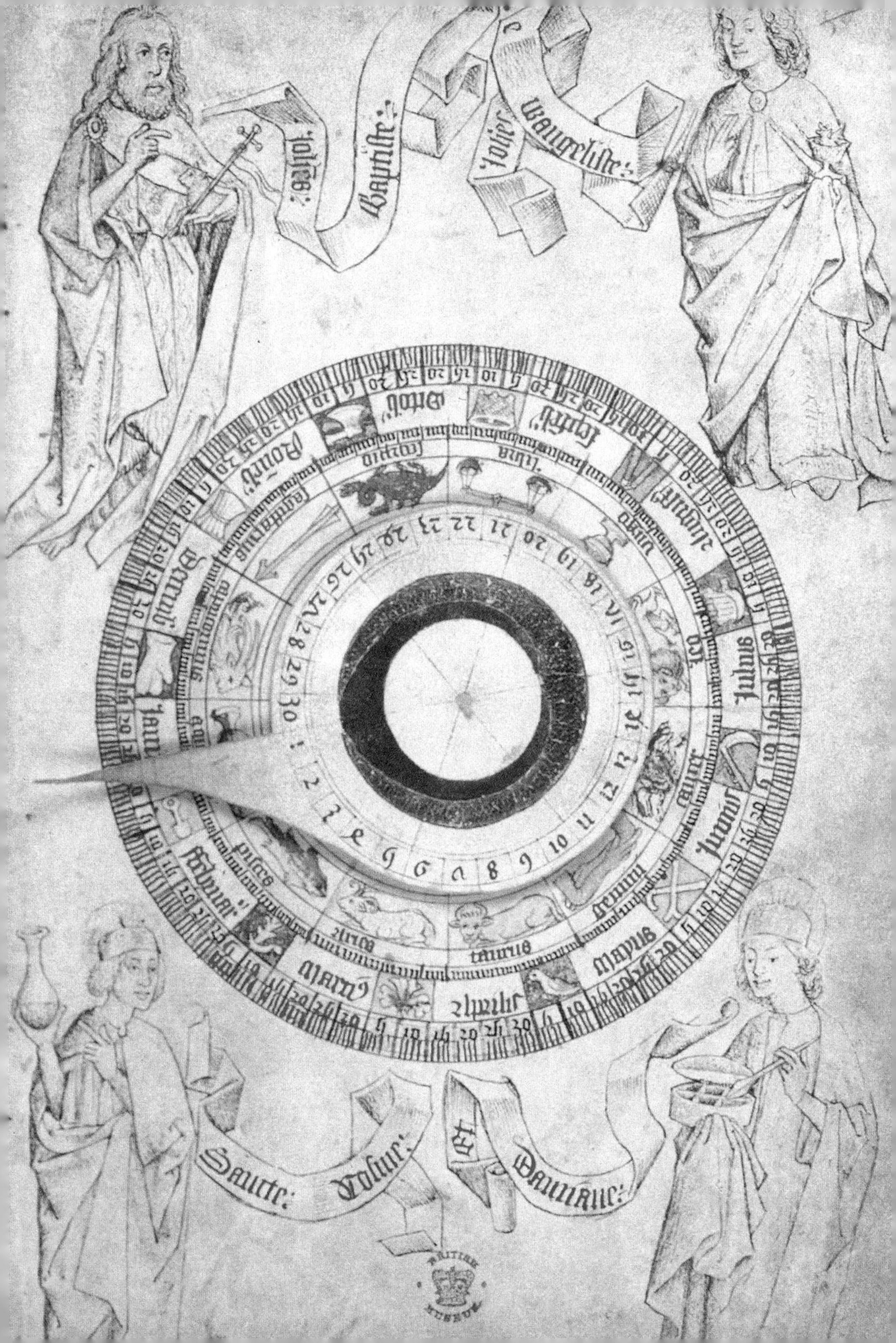

Johes: Baptiste: Johes Euangeliste:
Sancte: Cosme: Et Damiane:
BRITISH MUSEUM

More than 320 years later and known as *Old Moore's Almanack*, it is still a staple of British newsagents' counters, predicting the outcomes of world events, sporting fixtures and celebrity marriages, alongside tide tables and phases of the Moon for lunar gardeners and night fishermen. The first American almanac, *An Almanac for New England for the Year 1639*, began a slow trickle of imitations including one by none other than Benjamin Franklin, who, under the pseudonym "Richard Saunders", began *Poor Richard's Almanac* in 1732.

Gardening by the stars has long been practised – the Greeks considered the appearance of the Seven Sisters (Pleiades) constellation a fine time for hoeing weeds – but the craft evolved through the centuries into folk gardening. Thomas Hyll, in *The Profitable Arte of Gardening* (1563), exhorted gardeners to respect the stars, "whose beames of lighte" had the power to both bring to life and destroy tender seeds. Even relatively recently, the Pole Star (*Polaris*) heralded a good time to sow parsley (if indeed, there was ever a good time to sow parsley, see page 56). Even now, biodynamic farming – combining organic practices with spiritual and ecological concepts and often using astrological calendars and phases of the Moon – has become incredibly popular, especially in the multi-million-pound wine business. Gardening by the heavens, like the stars themselves, has proved cyclical by nature.

Right Sixteenth-century woodcut showing how each part of the body is governed by a different star sign.

Opposite Night jasmine (*Cestrum nocturnum*) from *Plantarum rariorum horti caesarei Schoenbrunnensis descriptiones et icones*, Nikolaus Jacquin, 1797–1804.

Cestrum suberosum.

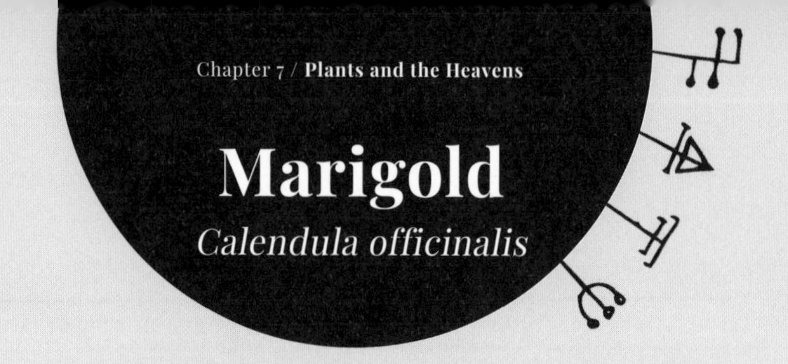

Marigold

Calendula officinalis

A classic pot-herb, marigolds have been used for flavour, dye and medicine for centuries.

The plant was named by the Romans, from the Latin *calendae*, meaning "little calendar", because it was said to flower on the first day of several months. So regular was the marigold's blooming that, in Wales, if it didn't open by seven in the morning, there would be thunder later.

Marigolds were cultivated in cottage gardens from medieval times onward and escapees still grow wild around inhabited areas. It was known as the "sunne's herb", "husbandsman's dial" and "summer's bride" because the flowers follow the Sun across the sky. The marigold is traditionally governed by the Sun and considered an element of fire.

Marigolds symbolize constancy in love. They are used in divination, simmered over a fire with wormwood (*Artemisia absinthium*), marjoram (*Origanum majorana*) and thyme (*Thymus*) in a concoction of honey and vinegar, then rubbed on the skin with various rhymes, bringing dreams of a future husband. Once snared, he could be kept by including marigolds in the bride's wedding posy.

The marigold's "golden" colour is remembered in myth (the flower was once a princess, accidentally touched by her father, King Midas) and in distant memory (in France, the circular golden flower was known as a "*gauche-fer*", the small iron shield medieval knights wore on their left arms). William Turner somewhat sniffily notes in *A New Herball* that some used marigold "to make their heyre yelow with the flour of this herbe not beyinge content with their natural colour which God hath given them". In the garden, marigolds acted as a sacrificial plant to keep pests from crops; in the kitchen, their petals decorated salads and added colour to stews.

By far the most important use for the flower, however, was in the medicine cabinet. The flower head, rubbed on a wasp sting, reduced the swelling. It was also used to ease ulcers and sores, fever and even plague. Toothache could be cured by rubbing marigold-steeped vinegar on the gums, while marigold tea treated measles (though some playground stories said the "measle flower" would give you the disease if you touched it). Hannah Woolley's *The Gentlewoman's Companion* (1675) suggested a conserve of marigold to treat depression. Today, calendula cream is widely available to reduce skin inflammations and nourish dry, chapped complexions.

Opposite Marigold (*Calendula officinalis*) from Köhler's *Medizinal Pflanzen*, 1897.

Compositae
(Calenduleae)

Calendula officinalis L.

The Wild Flora of Kew Gardens

Name: *Lunaria annua* L.

Vern. name: Honesty

Honesty

Lunaria annua

Honesty's associations with the Moon are present in its very name: *Lunaria.* Though not a night-flowering plant, its silver seed bracts glow in the light of the so-called "parish lantern".

Native of the Balkans and south-west Asia, the "moonflower" or "grandmother's spectacles" has lovely spires of white or purple flowers that grow to nearly a metre high in May and June, yet all its associated names are to do with their distinctive seed heads. It is "money in both pockets", "ready money", "shillings" and "silver pennies". In the United States, it is called "silver dollar"; in France, it becomes "*monnaie du pape*" ("pope's money"). Unsurprisingly, it is thought by many to bring good luck and wealth to anyone carrying a sprig in their pocket (in the Channel Islands, every bride hung a sprig in her new house to bring luck and happiness to her married life). Of course, with folklore's traditional dichotomy of one plant having two opposing meanings, other people won't have honesty in the garden, let alone the house.

It may be called "honesty" (from its translucent, three-layered seed-bracts that reveal everything inside them) but this plant has other, darker names: "the devil's ha'penny" and, in Denmark, "*judaspenge*" ("coins of Judas"). "Moonflower" will bring the bearer wealth, but it will be dirty money, ill-gotten. It was also one of the despised "thieves' plants", which could ward off demons and evil spirits because it was one of their own. Wielded by the dishonest, it could also open doors, break chains and unshoe horses.

Despite all this, honesty remained a surprisingly popular plant. Part of the Brassicaceae family, its seeds were occasionally used as a somewhat hotter mustard substitute. American colonial gardens grew honesty for its root. John Gerard calls the plant "bolbanac" or "white satin" and notes the seed is "hot and drie and sharpe of taste", while the roots are "somewhat of a biting qualitie, but not much: they are eaten with sallads as certaine other rootes are". The leaves, he thought, could make a useful unguent for wounds, and he had heard it recommended for "falling sickness" (epilepsy).

The plant is biennial and, once established in a garden, merrily self-seeds, yet somehow rarely becomes invasive – merely a beautiful ghost, luminous in the soft evening light.

Opposite Herbarium sheet of honesty (*Lunaria annua*), collected at Kew, 2009.

Chapter 8: Secrets of the Stillroom

Separate from other service areas, stillrooms were cool, dry chambers, locked and used only by the lady of the house, or a highly trusted servant. This was where the good stuff was kept: liqueurs and preserves, rare beverages such as coffee, tea and chocolate – and spices, including the family cone of sugar. It was also where household medicines were prepared.

Part storeroom, part kitchen, part laboratory, stillrooms were found throughout Europe, and early American settlers brought the concept with them as somewhere to create soaps and healing balms, syrups and tinctures.

Medieval stillrooms bore some similarities to the still-houses of alchemists, those mysterious natural philosophers who sought to purify base materials into precious metals. "Multiplication" – making gold out of thin air – had been outlawed in Britain in 1404 by King Henry IV and would not be repealed until 1689, but that didn't stop some folk trying it. A typical alchemical laboratory might include athanors (furnaces), grinding mills and vessels in which to distil curious substances. A regular (and entirely legal) household stillroom would have used a hearth, pestle and mortar, pots and pans. Alchemy pursued the philosopher's stone – the elixir of immortality and cure to any disease. Most women of the stillroom merely strove to keep their households in relative good health.

Here, herbs could be distilled into their essential oils for later addition to tinctures, salves, cleaning agents, pest repellents, flavourings and cordials. Methods were handed down from mother to daughter, sometimes in handwritten "receipt books", though by the sixteenth century, there was a burgeoning trade in published herbals and housekeeping manuals. Thomas Dawson's 1585 *The Good Huswifes Jewell* includes recipes for "approved medicines for sundry diseases" alongside pancakes, "sallets" and puddings. Admittedly, not all the ingredients were store-cupboard staples – one recipe for "courage" required the brains of three or four cock sparrows – but even then, most of the basic ingredients could be quickly found in the surrounding countryside or, even more handily, in a well-kept kitchen garden.

Herbs were gathered according to specially laid-out rules, often governed by the time of day, phase of the Moon and, more mundanely, the weather – they needed to be harvested on dry days. Roots were at their most succulent before any top growth had appeared; leaves and shoots were best just before flowering and ideally were used fresh. If they had to be stored, they were hung in bunches, upside down, in well-ventilated darkness, to preserve their colour and essential oils. When perfectly dry, they were placed in well-stoppered jars, also away from the light. Professional apothecaries had special drying sheds, but a dark corner of the stillroom sufficed for most.

Garden plants, and a few carefully purchased items like exotic spices, were the basic ingredients for tisanes, ointments, pomanders made of lavender (*Lavandula*) and the petals of sweet briar (*Rosa rubiginosa*), and made washing balls smell at least halfway decent. Sweetly perfumed waters had a wide range of uses, including strewing, medicines, perfumes, flavourings and the ever-needful handwashing during a meal, given few people had cutlery. Wealthy people could afford imported castile soap made from olive oil; everyone else made their own.

Lye soap had been made since Babylonian times. Ashes from the fire were soaked in water for several days to extract the potassium carbonate (potash). When the caustic concentration was strong enough

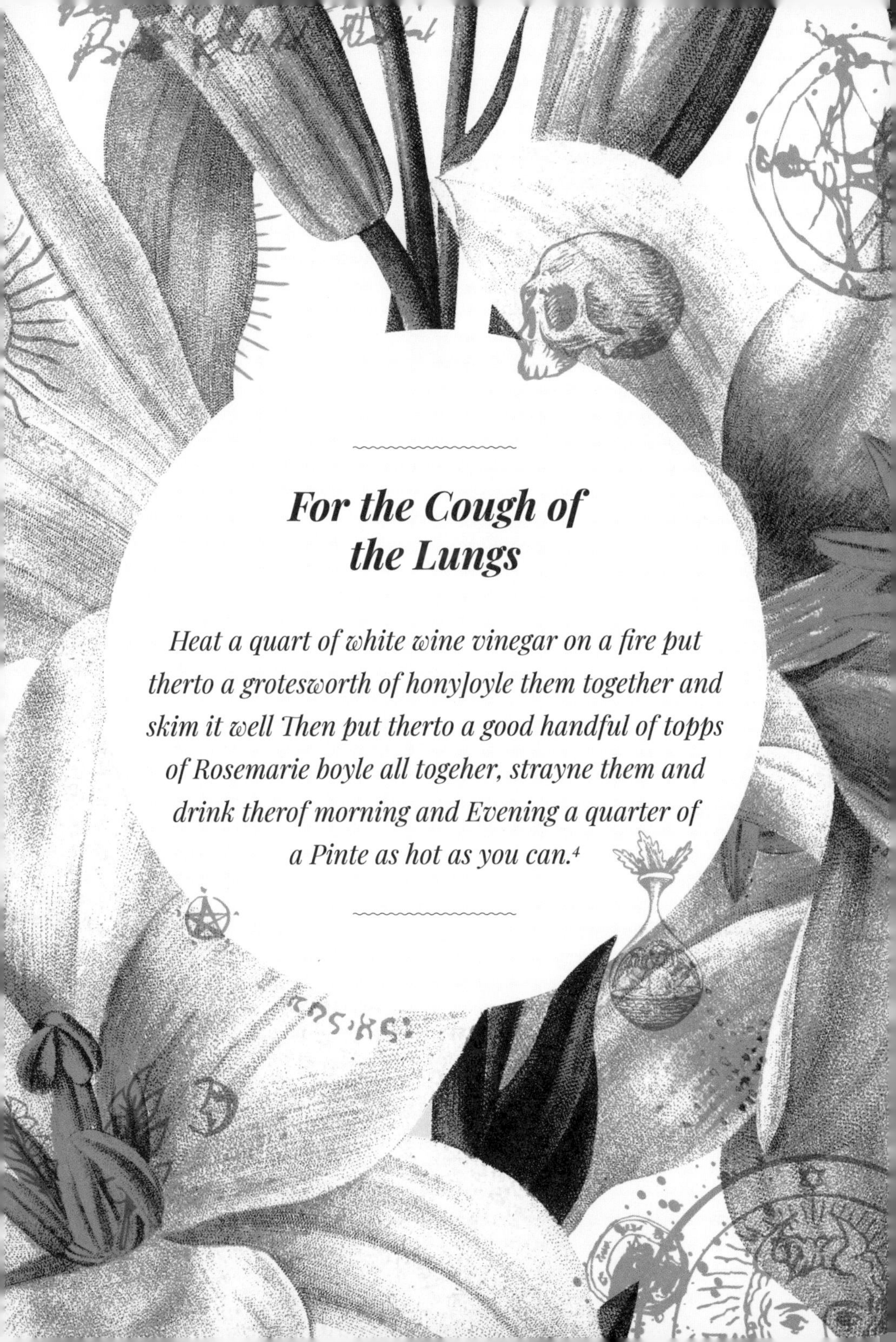

For the Cough of the Lungs

Heat a quart of white wine vinegar on a fire put therto a grotesworth of hony]oyle them together and skim it well Then put therto a good handful of topps of Rosemarie boyle all togeher, strayne them and drink therof morning and Evening a quarter of a Pinte as hot as you can.[4]

CITRUS MEDICA, LIMON — *Limone di giardino*

Pianta legnosa e quasi arborea perenne e di foglia sempre verde — Ama la terra forte ben concimata mista di terriccio vegetabile e vinacce ben macerate insieme — Si coltiva in vasi che in inverno si ripongono, a spalliere, o a boschetti che in inverno si cuoprono — Fiorisce quasi tutto l'anno ma specialmente in Marzo ed in Agosto; e matura i frutti un anno dopo nel mese di Maggio e Settembre — I frutti sono odorosi e contengono molto Agro —

to dissolve a goose feather (though not quite thick enough to balance an egg) the sludge was collected and boiled up with hog grease for washing clothes.

After lye soap, anything that could make the world smell sweeter was welcome. Scented herbs such as marjoram (*Origanum majorana*) and wild thyme (*Thymus*), and flowers like gillyflowers (carnations, *Dianthus caryophyllus*), woodbine (honeysuckle, *Lonicera periclymenum*) and jasmine (*Jasminum*) were grown close to the house so their scent could waft in through open casements, because even to smell them brought cleanliness and health. Bags made from dried flowers could soothe the brain, dispel melancholy and, of course, be added to that fancy castile soap to make it even lovelier.

Cosmetics, too, were born in the stillroom. May dew, gathered with a sponge, strained and exposed to sunlight, was a good general beautifier, though if gathered from the leaves of fennel (*Foeniculum vulgare*) or wood anemone (*Anemonoides nemorosa*) after night-time rain, it was particularly soothing for sore eyes. Wild strawberry leaves (*Fragaria vesca*), with cinquefoil (*Potentilla reptans*), tansy (*Tanacetum vulgare*) and plantain (*Plantago*) could be added to cows' milk that had been reduced to creamy thickness and kept up to a year for use as a face lotion. Freckles might be removed with distilled elder leaves (*Sambucus*).

Preparations made in the stillroom would keep the entire household healthy: master, servants and animals. General tonics, such as hippocras – a sugared, spiced wine – supposedly protected the wealthy and healthy from any number of illnesses.

Opposite Citron (*Citrus medica*) from *Raccolta di fiori frutti ed agrumi*, Antonio Targioni Tozzetti, 1825. Acidic perfume, clarifying oils and mouth-watering flavour make the many forms of citrus useful in everything from cough sweets to cleaning products.

Possets, made from wine, eggs, sugar and spices, were given to the poorly, elderly or infirm. Great care was taken in making this custard-like dish, which was traditionally poured from a great height to make the top foamy. Patients would eat the creamy "grace" with a spoon, then carry on digging into the smooth, semi-solid custard or "spoon-meat". If they were consuming the concoction from a proper posset cup, they could drink the last, highly alcoholic liquid base through its integral spout; failing that, they just upended the vessel and discarded the spoon before adding an eye injury to their woes.

Anchoring herbal preparations in solid or semi-solid form made dosages more accurate and remedies more stable and easier to administer. Ointments could be made by adding essential oils to beeswax, for example. Simples (medicines) were made fresh if possible, though some were created in advance, kept from spoiling with alcohol or sugar syrup. This made them expensive, especially in the days before plantations in the New World brought down sugar prices.

Medicated lozenges and pastilles have been used to soothe sore throats since the time of the ancient Egyptians, when lemon (*Citrus* × *limon*), herbs and spices were mixed with honey – the ingredients haven't changed much since. Sugar was easier to form into solids than honey. However, it was hideously expensive and time-consuming, as it needed to be nibbled from a large, moulded sugar cone with iron tongs, pounded into a powder and sieved of impurities before even starting the recipe. Before the invention of moulds, the boiled candy would have been "pulled" into a long rope of molten sugar and snipped into pieces with buttered scissors.

Cataplasms, or poultices, were moist compresses used to treat all manner of conditions.

Sores, ulcers, bruises, boils, ingrowing toenails, toothache, inflammation, stomach cramps and broken bones were all suitable candidates, in both humans and animals. Poultices consisted of some kind of binding agent, such as bread, linen or plaster, which would be soaked in various concoctions, according to the ailment. They usually brought heat to the affected area, being thoroughly warmed beforehand but also using naturally "hot" herbs such as horseradish (*Armoracia rusticana*). Linseed oil, from the seeds of the common flax (*Linum usitatissimum*), was particularly popular because the fibres within it expand when they come into contact with fluid, drying out sores and drawing foreign bodies such as thorns or splinters from a wound. Flax was extremely valuable, both medicinally as seed and in textiles, as linen fibre. A good harvest wasn't always guaranteed, however. Bells needed to be rung on Ascension Day, the fortieth after Easter, to herald the new crop and, perhaps harking back to earlier traditions, superstitious farmers also leapt over fires on Midsummer's Day just to be sure.

Strewing herbs were used throughout the house to keep floors clean, deter pests and bring fragrance to rooms. Those whose leaves produced aromatic oils when trodden on were particularly good, such as chamomile (*Matricaria chamomilla*), lemon balm (*Melissa officinalis*), sweet woodruff (*Galium odoratum*) and mint (*Mentha*). John Gerard claimed Queen Elizabeth I particularly favoured meadowsweet (*Filipendula ulmaria*) for her personal chambers, even though in some places, being creamy white, it was considered an unlucky plant to have in the house – indeed, it was said if someone fell asleep with meadowsweet in the room, they might not wake up. A compromise could be reached by strewing the lightly scented herb in churches for weddings, giving it another folk name: "bridewort".

Herbal medicines were extremely important in the early modern stillroom, but they couldn't necessarily be relied upon on their own. Many folk practitioners, especially in more humble homes unable to boast a special room dedicated to medicine, were not above invoking talismans and good-luck charms to guard against both illness and any evil spirits that might have their eye on the family. The humble herb

Above Engraving of meadowsweet (*Filipendula ulmaria*) from Gerard's *Herball*, 1633.

Opposite Frontispiece of *Le grimoire du Pape Honorius*, 1760, depicting a charm to protect livestock.

GARDE POUR LES MOUTONS,

Expliquée à la page 106.

No. 2

Publish'd as the Act directs by W. Curtis Botanic Garden, Lambeth Marsh 1786.

bennet (*Geum urbanum*), sacred to St Benedict, was generally blessed (though in some parts of the West Country was thought to encourage snakes) and made an excellent talisman to hang over the front door to prevent the devil entering the house, especially if woven with betony (*Betonica officinalis*). Rue (*Ruta graveolens*), also known as the "herb of grace", was hung at east-facing windows to ward off plague, which was generally thought by the British to be brought on contaminated winds from France. In the stables, animals, especially cattle, were protected with charms. Usually, this meant garlands of herbs, including ivy (*Hedera*), elder (*Sambucus*), St John's wort (*Hypericum perforatum*), rowan (*Sorbus*) and hawthorn (*Crataegus*). *Viburnum lantana* is commonly called the "wayfaring tree" because it tends to grow by country paths, but it is also known by some as the "coven tree". Even so, it was often planted around cattle sheds in the seventeenth century to repel witches.

for any quik thing in a mans eare

Drop juice of Rue, wormwood or of southernwood into the eare; to kill it.[5]

Stillrooms and their related skills gradually fell out of favour with a newly enlightened gentry, leaving the secrets of health and beauty to the care of poor relations, women of the village and professional apothecaries. Much was lost, or passed on incomplete, within today's folklore. Echinacea for colds, dock leaves (*Rumex*) for stings, marigold, or calendula (*Calendula officinalis*), for wounds. Children still sometimes learn to make a pomander for the wardrobe or at Christmastime by studding an orange with cloves, even if they don't know it would once have graced the garderobe (toilet), where clothes were kept, to repel pests. Some people still make lavender bags to line underwear drawers or to help them sleep. A few still infuse beeswax with herbs for use as lip salves and balms – we just tend to use a kitchen these days, instead of a stillroom.

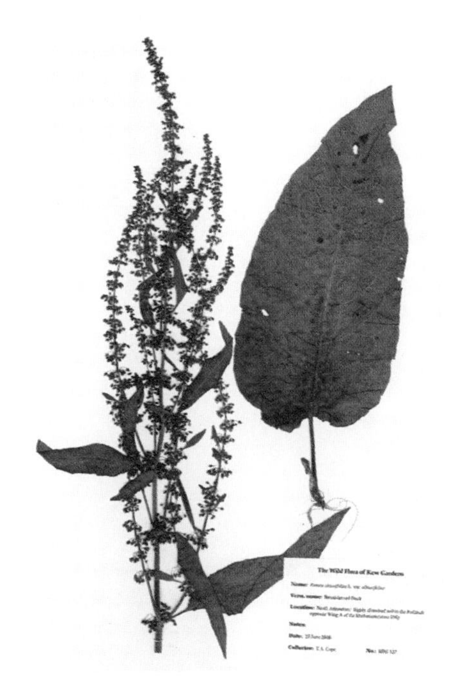

Above Herbarium sheet of broad-leaved dock (*Rumex obtusifolius*), collected at Kew, 2009.

Opposite Narrow-leaved purple coneflower (*Echinacea purpurea*) from *Curtis's Botanical Magazine*, 1787.

Rowan tree

Sorbus

Vibrant against a leaden late-autumn sky, the rowan's clusters of bright orange berries announce winter is on the way.

If the crop is heavy, the harvest will be as well, but it will also be followed by a harsh winter. For some, rowan is the "witches' tree", used for evil purposes. Most disagree, believing the plant to protect against evil. Often called the "mountain ash", rowan has no family connections to the ash tree; it is more closely related to apple, hawthorn and rose. "Lady of the mountain", "quicken tree", and "wildwood" can be found in some of the highest, most inaccessible places of Northern Europe. In Scandinavia, the hardiest examples, clinging to soil-less gullies between rocks, are known as "flying rowans". In Scotland, the breeze rushing through the feathery leaves lends the name "whispering tree' to the holder of secrets.

Hebe, the Greek goddess of youth, had her magical cup of ambrosia stolen by demons, so the gods sent an eagle to retrieve it. During a fight for the chalice, the eagle shed both blood and feathers, which turned to rowan trees as they hit the ground. In Norse mythology, rowan was sacred to the god Thor after it saved him from the fast-moving river of the Underworld by bending its branches as a lifeline. Celtic traditions tell how the first woman was formed from rowan (the first man was an ash). The feminine associations continued into Christianity. Irish St Brigid's name might come from Brid, the Celtic goddess of the hearth, arts, healing, midwifery, smiths and weavers. Her sacred plant is the rowan, which is why spindles and spinning wheels are traditionally made from rowan wood.

Each of the rowan's berries bears a tiny, five-pointed star at its base that, as the fruit ripens, looks like a magical pentagram, a sign of good luck. Rowan protected coffins and babies' cradles, and was attached to the horns of cattle. It is still thought unlucky to fell a rowan tree, especially in the Scottish Highlands. For some, rowan was the tree of the fairies. It was wise to keep a rowan staff handy at midsummer as an escape route, should someone become trapped in a fairy ring. It could also be used as a divining rod for metal ore.

Rowan has always been part of the traditional herbal medicine chest, believed to be effective as a purge for diarrhoea, a gargle for sore throats and an ointment for haemorrhoids. It was also purportedly a cure for scurvy, which makes sense – the berries are high in vitamin C.

Opposite Partly coloured drawing of a rowan tree, (*Cormus domestica*), by Mary Anne Stebbing, 1946.

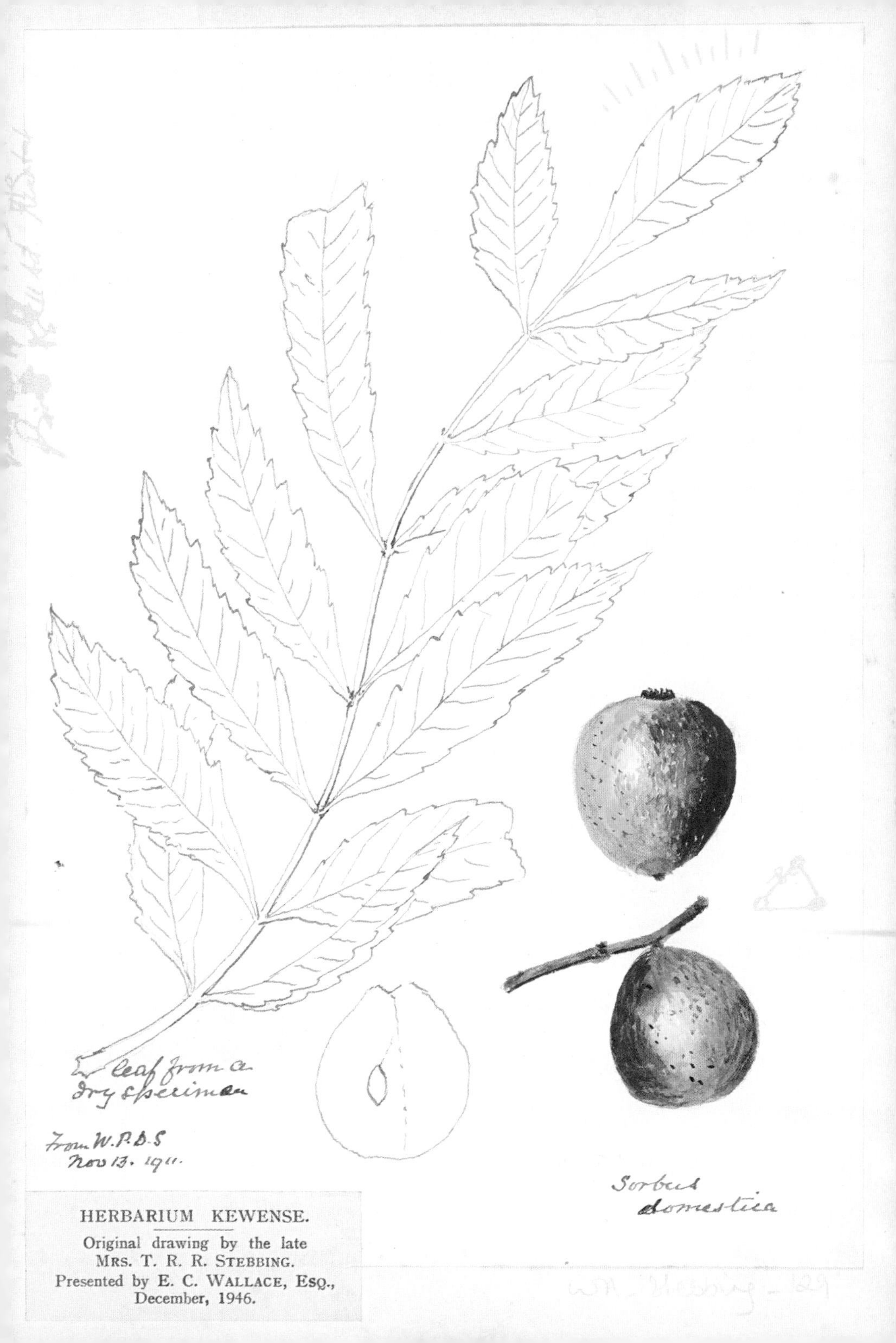
leaf from a
dry specimen
From W.P.D.S
Nov 13. 1911.
Sorbus
domestica
HERBARIUM KEWENSE.
Original drawing by the late
MRS. T. R. R. STEBBING.
Presented by E. C. WALLACE, ESQ.,
December, 1946.

Rosa Damascena od pallida Offic. Einfache rothe Zücker Rose, *Damascen* Rose.

Rose

Rosa

For many, the rose is the queen of flowers. Its perfume is legendary, its seductive qualities unparalleled.

The rose has been a symbol of England since Pliny the Elder couldn't decide whether the land was named Albion for its white cliffs or for the roses that grew wild on its waysides.

Cleopatra is said to have seduced Mark Antony on a bed of roses; the Empress Josephine collected every known variety at Malmaison, the château she shared with Napoleon Bonaparte. Greek mythology tells of Aphrodite pricking her foot on a thorn from a white rose bush, and of Cupid spilling some of his magic love potion. Both accidents stained the flowers red. It's said roses sprang up wherever St Alban trod on his final journey to martyrdom and, every 22 June, he is remembered in a service at St Albans Cathedral, in which the Bishop carries a rose-festooned crozier.

The annual Knollys Rose Ceremony in the City of London sees a "fine" of a single red rose presented to the Lord Mayor. The original "crime" – building a bridge between two adjoining properties without permission – was committed in 1381. Similarly, "peppercorn" rents are still occasionally levied today. Naomi House, a children's hospice in Hampshire, pays for its 99-year lease with one dozen red roses each Midsummer's Day.

Rose petals have been used since antiquity to garland a room or scatter over a newly married couple, but the act of giving roses is a minefield. An entire dialect of the Victorian "language of flowers" is devoted to roses. Some meanings are obvious: white is purity, red is passion and rosebud is innocence. Others are more complex – yellow can mean both jealousy and friendship, Rose de Mai (*Rosa* x *centifolia*) voluptuous love, and dog rose (*Rosa canina*) pleasure and pain. Roses were painted on the ceilings of Roman banqueting rooms in honour of Harpocrates, god of silence, reminding people to use discretion in their dinner talk. Matters discussed "under the rose" should remain "sub rosa".

Rose is a customary flavouring for *lokum* (Turkish delight), while sauce *eglantine* (from *Rosa rubiginosa*, or briar rose) was poured on Queen Victoria's desserts. Damask rose (*Rosa* x *damascena*) petals were added to cherry pies to deepen the flavour. Rose water is considered gentle and cleansing, even for sensitive areas such as eyes.

Culpeper found roses useful in preventing vomiting, haemorrhaging and "overflowing of the menses", but especially efficacious for colds and coughs. Indeed, rosehip syrup was popular well into the twentieth century and children were paid to collect the hips during the Second World War.

Opposite Damask rose (*Rosa* × *damascena*) from *Paxton's Flower Garden*, *c.* 1850.

Hawthorn

Crataegus monogyna

While the ancient Greeks garlanded newlyweds with hawthorn and lit the way to the bridal chamber with hawthorn torches, hawthorn is feared in many northern European traditions.

Sitting under a "hag-thorn" in May invited abduction by fairies. Even today, a single bush is often left in the middle of a field rather than risk a fairy's wrath at cutting it down.

When not alone, "quickthorn", named for its speed in establishing, was a useful hedging plant. In some regions, a specially cut branch was burned in each field on New Year's morning to encourage a good crop. The charred remains were preserved until the following year.

Hawthorn blossom was rarely welcomed in the house. The flowers were white, but it was also said to smell of death, or of the plague. There is a grain of truth in this, as "May blossom" contains trimethylamine, a chemical also present in decaying animal tissue. Hawthorn is another contender for Christ's crown of thorns, even though it blooms in May, the month sacred to the Virgin Mary – this fact mitigates it in some eyes, especially if the "whitethorn" is blessed with holy water.

Folklorists still disagree whether the old saying "Cast n'er a clout 'til May be out" means you must keep your coat on until the month of May is finished, or until the "May blossom" appears. "Bringing home the May" – pleasure jaunts to view hawthorn blossom – was a popular Tudor pastime; we have several accounts of King Henry VIII "going a-maying", meeting "Robin Hood" and "Maid Marion" along the way, King and Queen of the May. Samuel Pepys's wife Elizabeth bathed in hawthorn dew on a May morning in 1667.

Hawthorn wood was sometimes used for divining rods. It was also said to prevent lightning strikes and to burn the hottest of all firewood. New hawthorn leaves are sometimes known as "bread and cheese" because they were eaten in times of want. The fruits or "haws" have some delightful local names, including "haggle-berries" and "pixy-pears". A tincture from them was remedy for heart conditions while, beaten to a powder, they treated "the stone" (stone disease) and dropsy (oedema). Distilled thorn water was said to draw splinters from the skin and the bush also acted as "rag tree": people would tie slips of fabric to the bush, which would then bear the weight of an illness.

Legend holds that Joseph of Arimathea created the Holy Thorn at Glastonbury by planting his staff in the ground, after which it sprang to life.

Opposite Hawthorn (*Crataegus monogyna*) from *Flora Danica, c.* 1761–1883.

Caprifoliaceae.
Sambucus nigra L.

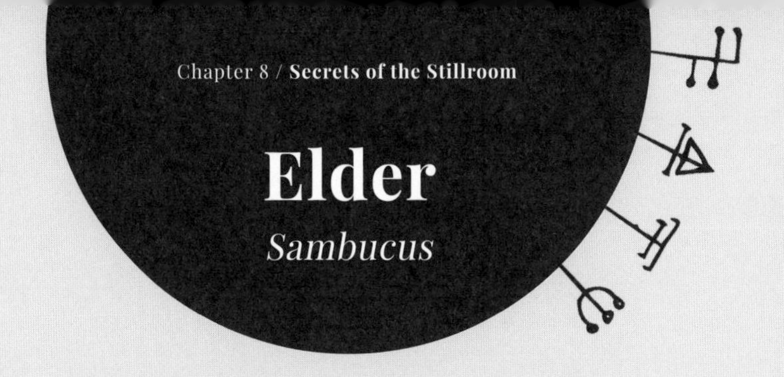

Elder

Sambucus

Legend holds that Judas hanged himself in an elder tree, not boding well for its reputation in Christian traditions, especially when it was said that Christ's cross was made from elder wood.

The early Church seemed very keen to decry "the tree of death". Once a fine forest tree, the elder was said to have been cursed into a malformed shadow for its part in the Crucifixion, its magnificent berries shrunken to clusters of tiny black balls. It stank from the bodies of criminals; invisible witches and fairies rode the branches that switched about in the wind. Anyone sitting under an elder on Midsummer's Eve would see the fairy king and his retinue passing, and if they fell asleep would never wake up. Elder pith dipped in oil and lit, floating in a glass of water on Christmas Eve, would reveal all the witches in the neighbourhood.

Of course, the Church had an agenda. It needed to obliterate earlier, older mythologies that revered the elder as one of the most important trees in the sacred grove – mythologies where the elder was only "bad" if treated without respect. Sacred to the Northern European goddess Holda, the elder could ward off evil spirits, something that eventually did settle into the Christian Church. Crosses made from elder wood were hung inside stables and travellers carried an elder twig to ward off thieves. Sailors would carry twigs, too; their relatives could judge their fate by the health of the tree. Elder amulets warded off St Anthony's fire (erysipelas or ergotism), rheumatism and epilepsy. Self-sown elder was best, as it deflected lightning. Elders should never be planted near wells, but were often found near privies, as they love rich damp soil. Their foliage also contains a natural insect repellent.

The only serious mistake one could make with an elder was to cut it. This was stealing from Holda, punishable by death within three days. If you had to cut the tree, she had to be asked politely – your hat removed and knees bent – with the reason for needing the wood clearly stated and a promise to repay in the future. As firewood, elder screams with pain when burnt; the boiling sap is said to be the devil spitting down the chimney.

Pliny called elder "pipe tree" because of its hollow stems, which were useful in making fifes, trumpets, penny whistles and bagpipe chanters. As medicine, the tree it was invaluable. In 1655, Thomas Browne suggested it for quinsy (mouth abscess), sore throats and "strangulation". Culpeper believed elder leaves, stuffed up the nostrils, purged the "tunicles of the brain".

Opposite Elder (*Sambucus nigra*) from Köhler's *Medizinal Pflanzen*, 1887.

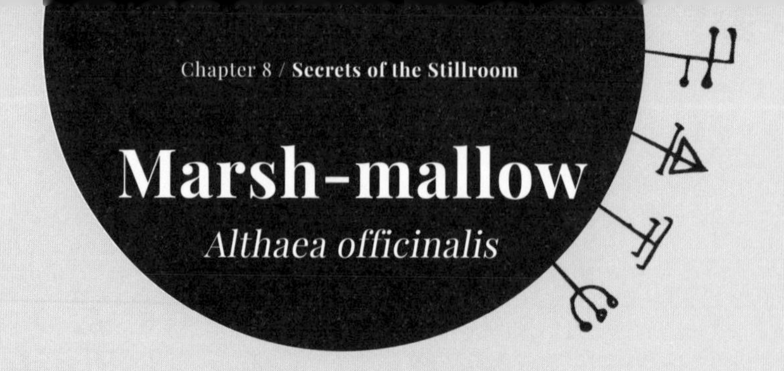

Marsh-mallow

Althaea officinalis

There are over 1,000 species in the Malvaceae family, but the most famous has to be the marsh-mallow, whose very name, *Althaea officinalis*, means healing, from the Greek *altho*, "to cure".

Originally from China, marsh-mallow was known to the Egyptians, Syrians and Greeks. The Romans stuffed suckling pig with the herb. It was used as a laxative, while mallow tea calmed internal inflammation and the leaves made a compress for sore eyes or a poultice for poorly horse legs.

British children also gathered mallow seeds, which form little roundels, giving the plant many local names: "nutlets", "flibberty-gibbet" and the mildly disturbing "frog cheese". The fleeting, purple flowers, decorating many a wayside or scrap of waste ground, lend yet more names, from "wild geranium" to "rags and tatters". This healing pot-herb had many virtues and few, if any, vices. Pressed into the mouth, the leaves were believed to relieve toothache and eased sore throats and coughs. Dried or fresh, they made a decent clyster (enema). Mashed, the plant helped sprains and stiff joints. Culpeper thought the roots, boiled in wine, good for "ruptures, cramps or convulsions of the sinews". Sweetened with syrup of violets, marsh-mallow cured painful urination, though potential Lotharios needed to beware – it could also act as an anti-aphrodisiac.

In Irish folklore, mallow is one of the seven plants that nothing evil can harm (the other six are St John's wort, eyebright, vervain, speedwell, yarrow and self-heal.). It would be most efficacious if gathered on a bright day near full moon, but if gathered on May Eve, while invoking Satan, it could be dangerous. Boys in County Limerick took great delight in hitting passers-by with the herb that day, as "protection" from such evil.

Marsh-mallow's anti-witchcraft properties might be temporary, and its roots, boiled with raisins and drunk in the early morning, were only believed to protect against disease for a single day, but it is a well-meaning plant. And if the odd medieval suspected villain rubbed a paste of marsh-mallow and egg white on their hands before undergoing the traditional trial-by-ordeal of holding a red-hot iron so they might hold the rod for a tiny bit longer and prove their innocence, who could blame them?

Mallow has mucilage in its fibres that thickens in water. In olden days, heated with sugar, this made a sweet "marshmallow" paste, used in sweets and to bind medicinal pills. Alas, the only thing in common with the puffy fireside candies that bear the name today is the sugar.

Opposite Marsh-mallow (*Althaea officinalis*).

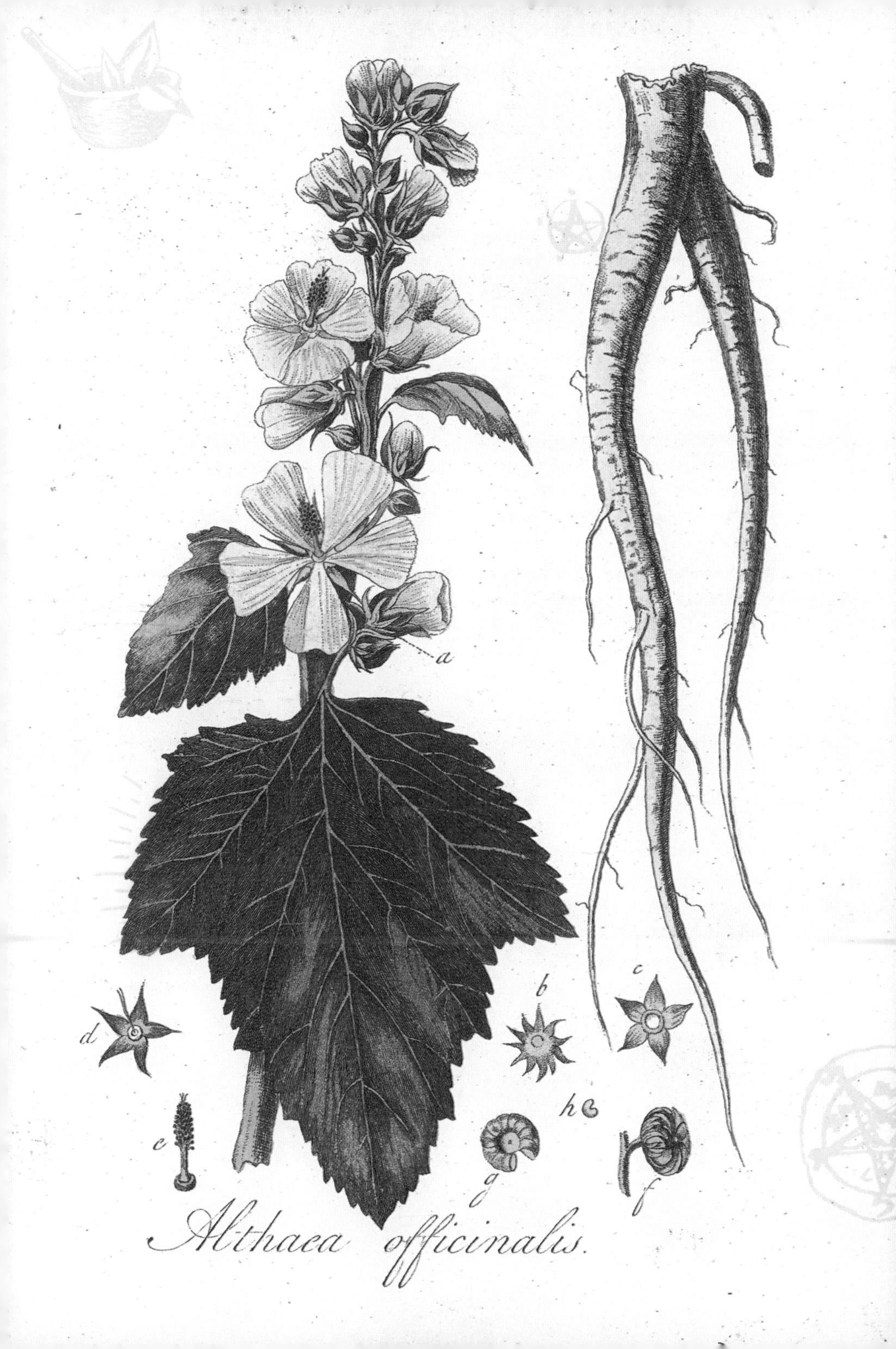
a
b
c
d
e
h
g
f
Althaea officinalis.

VERBENACEÆ
Verbena officinalis (L.)
Vervain
NATURAL ORDER Verbenaceae
DATE July 14th 1895
HABITAT Field border of Folkestone

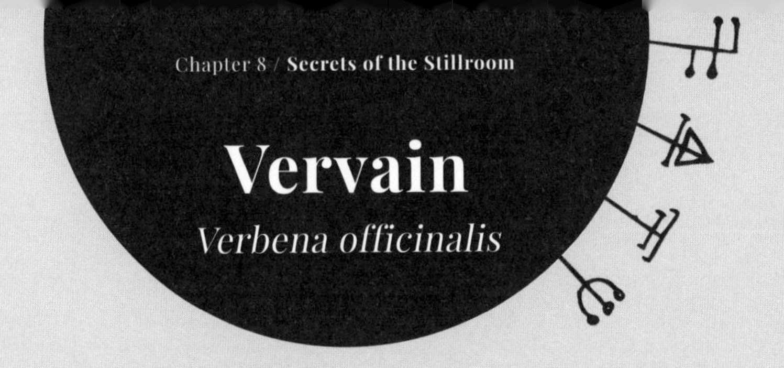

Vervain

Verbena officinalis

Vervain, also known as verbena, has held a potent place in folklore and medicine for millennia. But for all its magical claims, in its wild form, it's a scrawny individual.

Usually found on chalky soils and scrubland, its tiny, lilac-coloured flowers peep from coarse spikes poking between sparse, lobed leaves.

The Egyptians called verbena's dainty flowers "tears of Isis", though the Greeks and Romans claimed Hera/Juno as the weeping goddess, reflected in another common name for the plant, "Juno's tears". Pliny the Elder said verbena (Latin for "sacred bough") was introduced to Rome by the Celts, who cut "*ferfain*" ("witch's flower") with an iron blade at the rising of Sirius, the Dog Star. The Romans used verbena in divination and casting lots, even celebrating it with a festival called Verbenalia. The ancient Persians considered vervain an aphrodisiac, though this probably has nothing to do with the nineteenth-century European tradition of including the plant in a bride's wedding posy.

Vervain acquired a reputation as a veritable cure-all. Saxons believed it protected against thunderstorms. The Aztecs thought its root a diuretic, while Native North Americans used it against insomnia, circulation problems and headaches. Even the Christian Church didn't mind verbena, declaring it the "herb of grace".

By the Middle Ages, verbena was the "simpler's joy". It could cure the plague, gout and piles; gather doves; ward off witches and demons; drive away snakes; protect babies; magically sharpen blades and prevent dreams. In some places, it was known as "pigeon grass" because it was said birds ate it to improve their eyesight. A small black silk bag of verbena leaves, hung around a weakly child's neck, averted infection.

The "divine weed" could repel rabid dogs and deflect sorcery or witchcraft, but it did have a darker side and was said by some to be used by witches. Another "thieves' plant", vervain was boiled with rue and gun flints to ensure they fired, and it was said that if someone made a small cut in their hand and stayed the blood with a leaf of verbena, locks would magically open for them.

Even as late as 1837, *London Pharmacopoeia* claimed that a necklace of vervain root tied with a yard of white satin would ward off the king's evil (scrofula). It is still believed useful in combating anxiety, soothing the mind and promoting sleep.

Opposite Herbarium sheet of vervain (*Verbena officinalis*) collected in the UK, 1895.

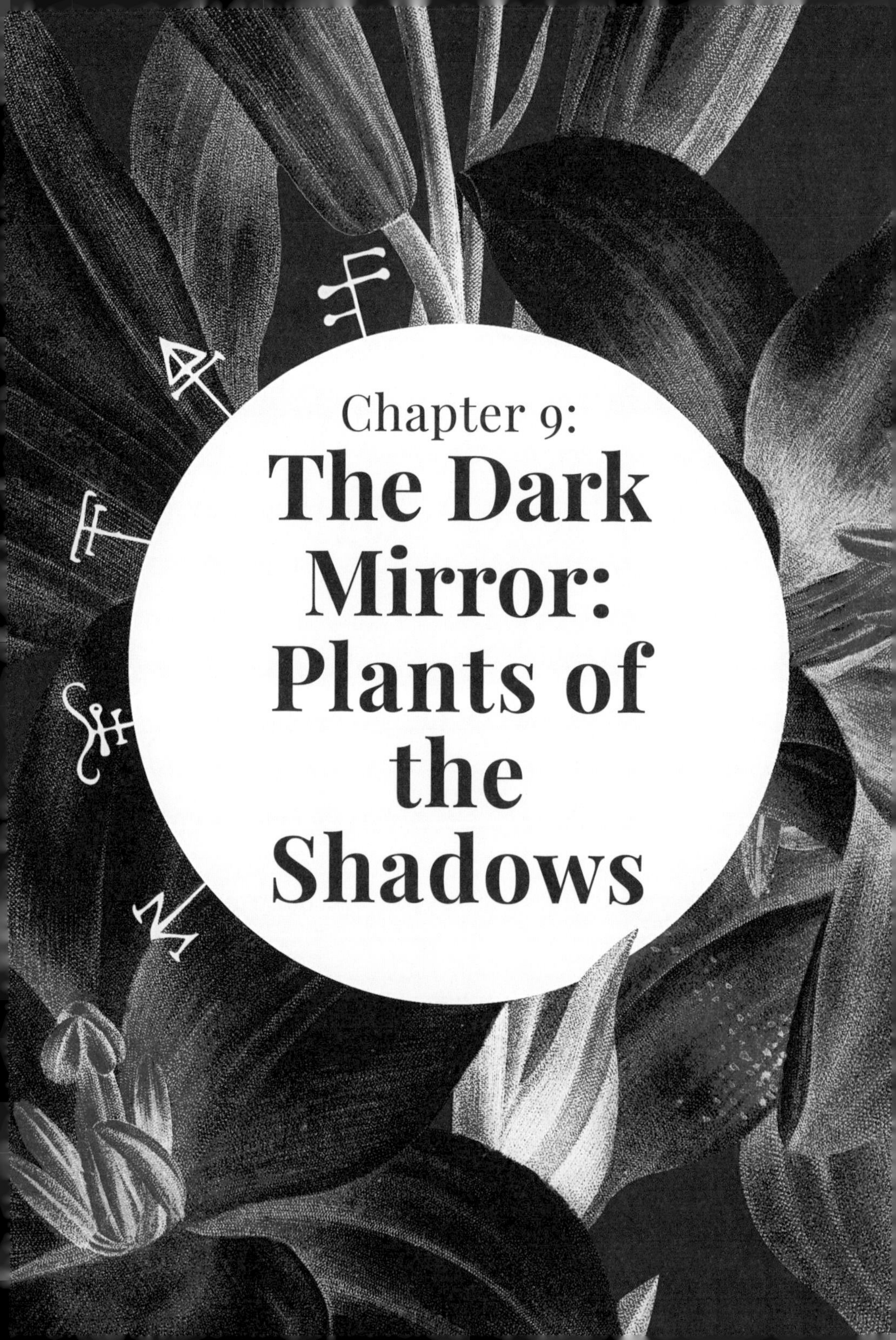

Chapter 9: The Dark Mirror: Plants of the Shadows

There was no doubt in the ancient Roman mind that magic could be sinister. The Latin word for "sorceress" was *venefica*: "female poisoner". The dark side of magic has always been with us, as have the "evil" herbs, plants that even people who "don't know plant names" fear: henbane, monkshood, deadly nightshade, foxglove, hemlock, mandrake. The great herbal poisons are infamous – but they're not alone. Our hedgerows and woodlands are filled with plants of the shadows. Used with the greatest of care, some can be useful. Few know them well enough, however, to try.

One of the biggest problems in tracing herbs used in spell-casting is that, after several thousand years, no one can quite agree which plants are protective, which will bring luck and which will cast boils on your next-door neighbour.

Some are attributed with directly opposing qualities, both good and evil; others are merely facets of the same quality. To medieval plague sufferers, rue (*Ruta graveolens*) was the protective "herb of grace", yet it became corrupted when it was used in "the vinegar of the four thieves". A group of robbers planning to strip plague victims of their worldly goods covered their noses with handkerchiefs doused in the concoction, which also included copious amounts of garlic and other pungent herbs, to prevent catching anything nasty themselves. It's possible the potion worked by repelling disease-carrying insects. It is said that when the gang was finally arrested, they were given their freedom in return for the recipe. Many variations of that bug repellent still exist today.

Dark herbs are often found in woodlands. The forest floor, especially in springtime, is a riot of flowers, as the sun shines for a few brief weeks before the canopy grows. Areas around streams are particularly fruitful. Ancient "coppiced" woods, where the young shoots of trees were harvested down to their stumps every so often, left temporary gaps in the trees, encouraging even wider varieties. The coppiced "stools" were tiny micro-worlds in themselves, with mosses, lichens and other plants growing between the cut shoots. Moss, in European folklore, was the microscopic realm of the mysterious "moss folk" – wise, aged and not to be trusted. The original "borrowers", the moss people could change size and shape and rewarded simple offerings of milk or porridge by healing animals or leaving gifts. Quick to offend, however, they punished rudeness or neglect with stealing anything from farm implements to children.

As forests are slowly carved up by human development, the natural "green corridors" that once allowed animals carrying seeds, spores and fruits from one copse to the next are disappearing. Plants, like the creatures that spread them, become trapped; some disappear altogether. The ghost orchid (*Epipogium aphyllum*) is one of the UK's rarest plants; it lives in the darkest shadows, needing neither leaves nor chlorophyll. Its pale, ethereal flowers may take 30 years to appear. Another rarity, spiked rampion (*Phyteuma spicatum*) was used as a medicinal herb for centuries, but has suffered since humans stopped coppicing woodland. Both are critically endangered.

The suffix "-mancy" comes from the ancient Greek word *manteía*, referring to divination. Since time immemorial, humans have tried putting all sorts of prefixes to it, from "pyromancy" (fire) to the rather more sinister "necromancy" (death), in an attempt to see into the future. Tea-leaf reading (tasseomancy) is unlikely to kill the scryer, but it's

Opposite Foxglove (*Digitalis purpurea*), from Köhler's *Medizinal Pflanzen*, 1887.

Scophulariaceae.
7
2
5
7
3
4
6
8
9
B
A
10
11
12
13
W. Müller n. d. Nat.
Digitalis purpurea L.

been suggested that priestesses at the Delphic Oracle inhaled fumes from the noxious herb henbane (*Hyoscyamus niger*) to give themselves visions of the future.

Botanomancy uses plants to conjure privileged information. It often involves the burning of leaves, herbs or branches (sometimes a specific question might be carved into a log or leaf first), listening to the crackling, staring into the smoke and inspecting the ashes for answers. Some diviners also interpret patterns formed by roots or branches. Rubbing crushed herbs between the hands could yield insights, as might "anemoscopy" (from the Greek word for wind) – writing a question on leaves and throwing them to the breeze.

Vervain (*Verbena officinalis*) has always been one of the most popular plants burned in botanomancy, though fig (*Ficus*) and sycamore (*Acer pseudoplatanus*) were potentially useful, in which case the process was known as "sycomancy". Other useful plants included lavender (*Lavandula*), mint (*Mentha*), mugwort (*Artemisia vulgaris*), yarrow (*Achillea millefolium*), sage (*Salvia officinalis*), marigold (*Calendula officinalis*), coltsfoot (*Tussilago farfara*), dandelion (*Taraxacum officinale*) and white willow (*Salix alba*).

It's possible that William Shakespeare's audiences weren't as horrified by his most famous divination brew as we might be today. When the Weird Sisters reveal to Macbeth their ingredients,

Above right Adder's tongue (*Ophioglossum vulgatum*) from William Burchell's *Saint Helena Journal*, 1806–10.

Opposite A witches' coven as imagined by Hans Baldung Grien in 1480. Sky-clad hags ride goats, brew potions and cast hexes while their feline familiar keeps guard.

"Eye of newt and toe of frog Wool of bat and tongue of dog Adder's fork and blind-worm's sting," they may simply be referring to mustard seeds (*Brassica nigra*), buttercup (*Ranunculus acris*), hounds-tongue (*Cynoglossum officinale*) and adder's tongue (*Ophioglossum vulgatum*). Admittedly, "Wool of bat" might have been trickier to source.

Curses were more serious stuff. Also known as jinxes and hexes, these dark spells were intended to do harm in return for a real or imagined slight. Some of the oldest curses we know are found on ancient Egyptian tombs, threatening misery to would-be grave robbers. Anyone could curse anyone; hexing wasn't just the domain of witches. Some illuminating curses can be found on lead tablets inscribed by ordinary ancient Romans – with the awful things they wished upon robbers, cheaters and neighbours – then left in wells, baths and temple walls for the gods to find. Biblical curses often punished the wrongdoer's heirs. Medieval "book curses" threatened ill luck on anyone stealing from or defacing the library, but outside the confines

of the monastery, hexes were more likely to dry the milk from someone's cow or make their horse lame.

It was said that, once invoked, curses could not be reversed, only deflected – or reflected back at the caster. Although most curses were spoken or written, plants sometimes helped them along the way. A blackthorn (*Prunus spinosa*) staff, pointed at a pregnant woman, would cause immediate miscarriage. In Ireland, if you wanted to jinx someone's crops, you buried something once alive but now dead, like wood ashes, an egg or a boiled potato, in the soil. In some places, asafoetida (*Narthex asafoetida*) was burned in a curse to force someone to leave someone else alone. Given its folk name "devil's incense", it probably kept everyone else away too. Some claimed that hanging a piece of seaweed by the privy cursed someone's urinary tract. It was not called "bladder wrack" (*Fucus vesiculosus*) for nothing: in the muddled world of folklore, it was also sometimes used to ease an overactive bladder.

Poppets – magical dolls – were sometimes used to cast spells, beneficial or malign, on a specific individual. Stuffing a poppet with knotweed (*Persicaria maculosa*), for example, was sometimes said to cause someone intestinal pain, though others claimed that all it did was control them or change their minds over something.

Wands, hand-held devices to channel power, are first mentioned by the Greek author Homer. In his epic *The Iliad*, the god Hermes uses a magical rod to cast sleep on humans, then reawaken them. *The Odyssey* has two magic sticks, one wielded by Athena to turn Odysseus into an old man, the other by the sorceress Circe to turn his men into swine. It is said the Druids used yew (*Taxus baccata*), hawthorn (*Crataegus monogyna*) or rowan (*Sorbus*) staffs but, by the thirteenth century, hazel (*Corylus avellana*) was recommended by *The Sworn Book of Honorius, a grimoire* (book of magic) of the time. For maximum quality, the branches should be new growth, cut with a single stroke at sunrise. Wands, and their big brothers, the magical rod and staff, could be any kind of wood, chosen for their various properties – for example, elder (*Sambucus*) could locate treasure, while willow (*Salix*) repelled evil.

Above left The Rebis, or "divine hermaphrodite" is a common image in alchemical illustration. It represents the ultimate "end result": both man and woman, sun and moon.

Opposite Ghost orchid (*Epipogium aphyllum*) by Walter Hood Fitch, from *Curtis's Botanical Magazine*, 1854.

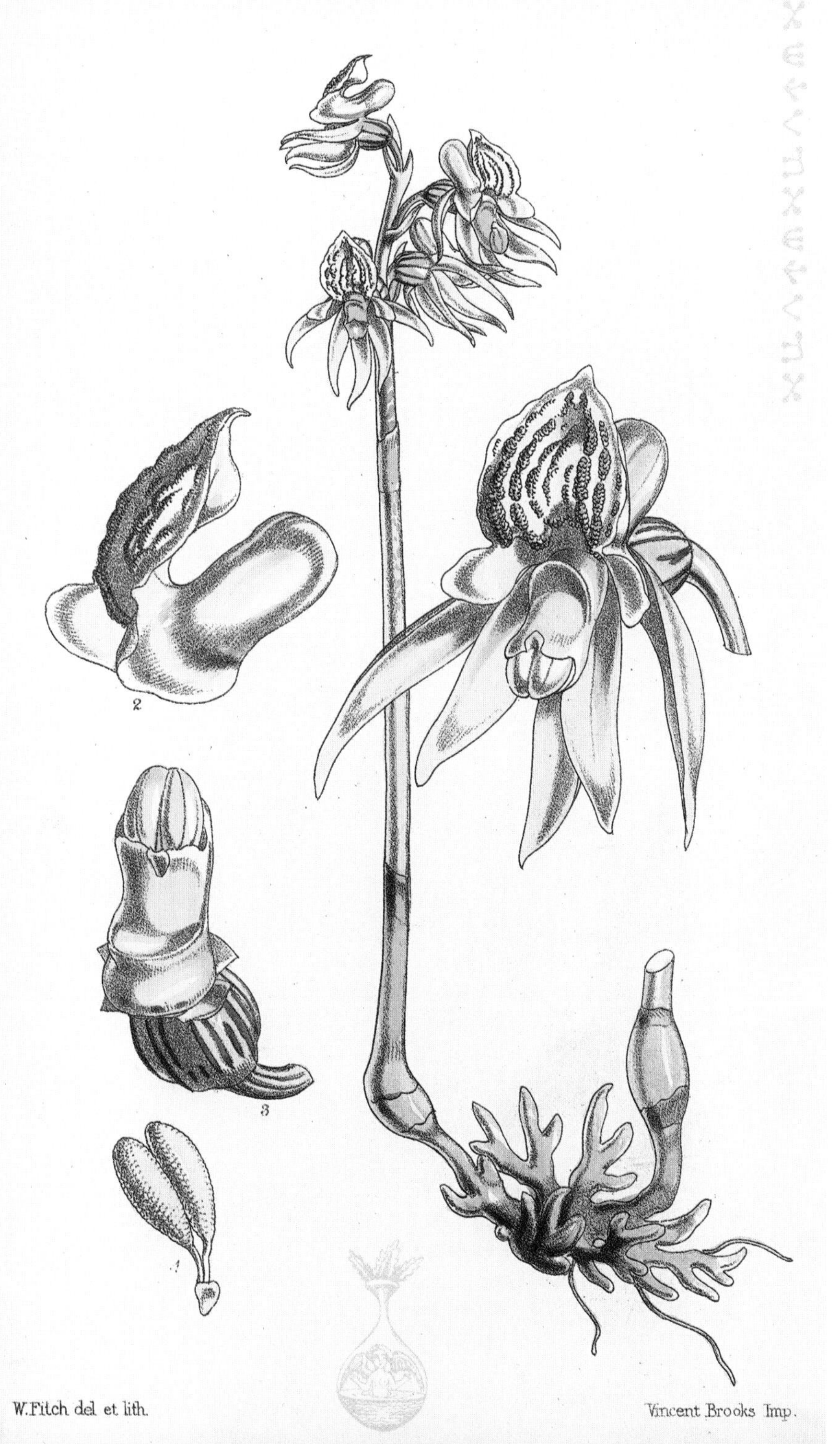
4821.
2
3
1
W.Fitch del et lith.
Vincent Brooks Imp.

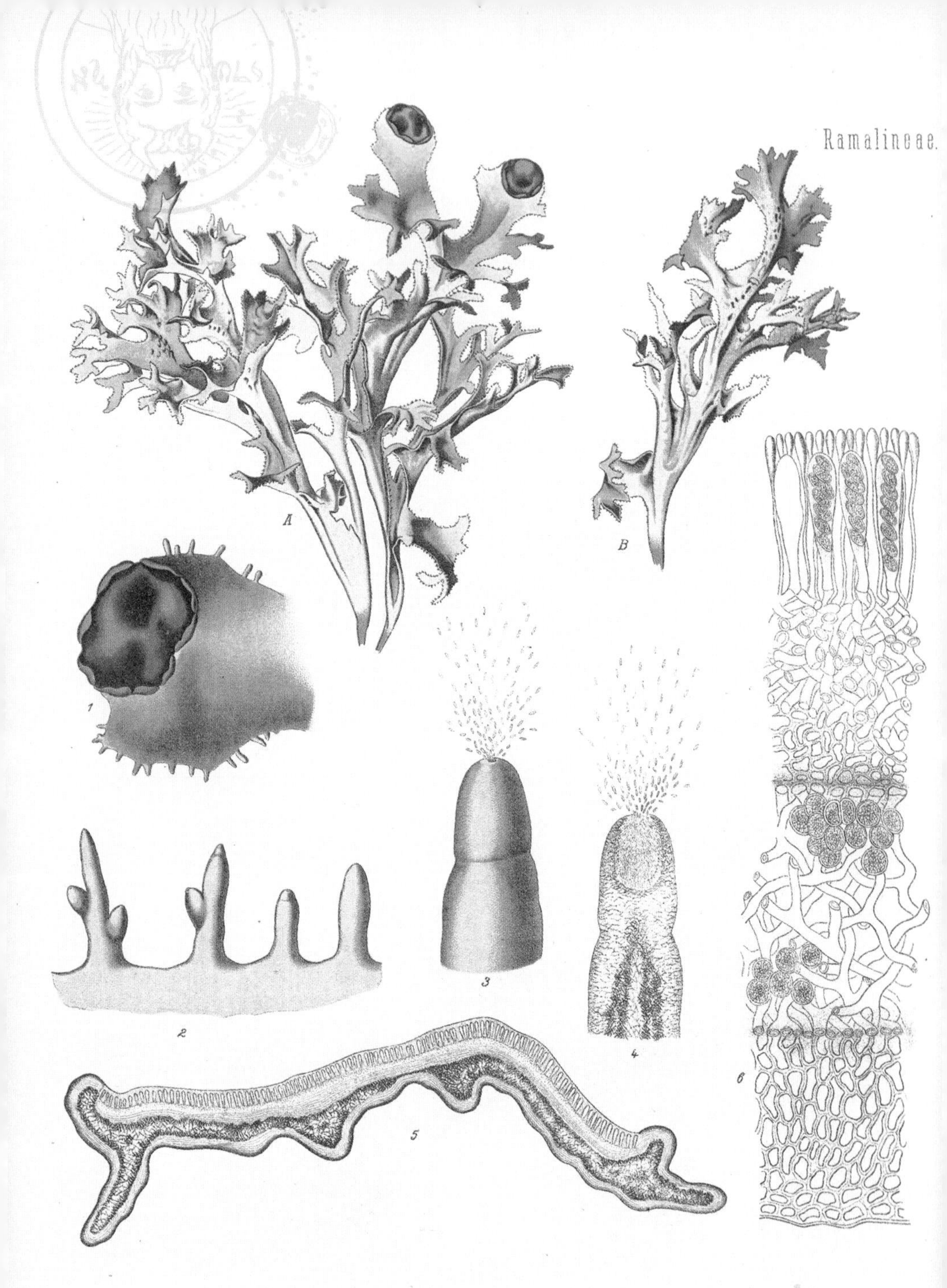

Cetraria islandica Acharius.

Only the very brave chose cypress (*Cupressus*), the tree of death, with which, it was whispered, one could communicate with Satan himself.

Although a wand was not considered innately "magic" – merely a tool through which power was channelled – to Christians already terrified by the idea of witchcraft, the stick was as dangerous as its wielder. By the Middle Ages, they had decided: witches were evil and needed to be rooted out.

Witchcraft, they said, left its mark everywhere. Wormwood (*Artemisia absinthium*) was used in pacts with the devil. Groundsel (*Senecio vulgaris*) sprang up where a witch had urinated. Plants that grew where blood had been shed were, themselves, flushed with red streaks. Protection was needed. Iron was useful – nails in trees and broken ploughshares buried under new fruit trees warded off evil – and no witch would step over salt. Any plants from the Cruciferae (now Brassicaceae) family were safe because their petals formed a cross, but white plants were nearly always unlucky to Christians, possibly because pagan traditions associate the colour with the white goddess. Thorn apple (*Datura stramonium*), also known as the "devil's snare" was particularly feared. Its night-blooming, trumpet-shaped flowers were sinister enough, but the plant's real power was concentrated in its highly toxic seeds. A brew made from thorn apple caused confusion and crazed behaviour, followed by deep sleep. Another thieves' plant, it was allegedly used to render victims helpless, then fog their memories after being robbed.

Opposite Icelandic moss (*Cetraria islandica*) from Köhler's *Medizinal Pflanzen*, 1887. A lichen that was once used across Scandinavia for making bread and in various medical treatments.

The stave (handle) of the witches' besom broom is made from ash for protection. Birch twig brushes, (to counteract evil spirits) are bound onto it with willow, honouring the goddess Hecate. It was used to clean ritual spaces of negative energy, but the Church couldn't believe that was all there was to it.

The idea of witches "flying" took off in the Middle Ages. The Church was highly suspicious of anyone they didn't have direct control over, and paranoia began to infiltrate common sense, whipped up by a little propaganda. The first image we have of witches flying on broomsticks is in a 1451 illustration, *Hexenflug der Vaudoises* ("Flight of the Witches") in the poem "Le Champion des Dames" ("Ladies' Champion"). The work was satirical but, two years later, Guillaume Edelin, Prior of St-Germain-en-Laye, near Paris, confessed, under torture, to having flown on a broomstick. He was condemned to death as a witch; it appears he died in prison before execution.

Flying wasn't necessarily literal. Even at the time, "flying ointment" was the thing people (especially men) really feared. This was a paste of hallucinogenic herbs, smeared on a broomstick and then rubbed on the genitals (and other mucus membranes, such as the armpits) to create a sensation of flying. In the early 1500s, Spanish physician Andrés de Laguna ran a "scientific" test with such an ointment, containing the usual suspects: hemlock, nightshade, henbane and mandrake. He covered the local executioner's wife from head to foot with the substance (far more than a regular witch would have been accused of taking) and then watched as she fell into a 36-hour sleep. She was not pleased to be awakened, claiming to have been surrounded by all the delights in the world, and to have cuckolded her husband with a younger, lustier lover. This did nothing to calm fears.

Sometimes strange behaviour happened even without broomsticks – which was, naturally, blamed on black magic. Neither animal nor vegetable, the curious fruits of fungi are, to the unskilled eye, a Russian roulette of food and poison. Some poisonous toadstools are indistinguishable from otherwise tasty mushrooms; others can't be seen at all.

We now know the ailment St Anthony's fire is caused by ergot (*Claviceps purpurea*), a fungus that grows on rye grass (*Secale cereale*), often in a wet spring following a cold winter. If the fungus was ingested unwittingly with the rye, the results could be very serious. Patients' faces turned bright red. They suffered hallucinations, a burning sensation, vomiting and even a strange, dry form of gangrene. Ergot poisoning has sometimes been blamed for the hysteria that led to the seventeenth-century Salem witch trials in the United States. It was said to be cured by pilgrimage to the Order of Hospitallers of St Anthony in Grenoble, an area in which ergot was not known. In the early twentieth century, scientists started seriously investigating ergot and, in 1932, came up with the alkaloid ergometrine, which is used in powerful modern drugs. There is still much to be learned from the dark garden.

Below Witches on broomsticks, consorting with the devil, from *Wonders of the Invisible Worlds*, 1693, by Puritan minister Cotton Mather.

"... a jar half-filled with a certain green unguent ... with which they were anointing themselves ... was composed of herbs ... which are hemlock, nightshade, henbane, and mandrake."[6]

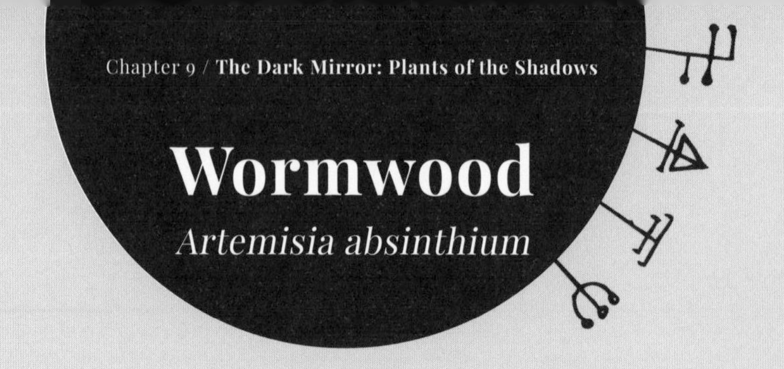

Wormwood

Artemisia absinthium

The mere word "absinthe" invites visions of wasted nineteenth-century artists languishing in Parisian gutters, lost in the throes of a stupor induced by the "green-fairy".

Manet's famous 1859 painting *The Absinthe Drinker* sums up a miserable life: drab browns and blacks, lightened only by the exotic, sparkling crystal bottle.

Beaten only by rue (*Ruta graveolens*) for bitterness, wormwood has always had an edgy reputation. Biblical tradition had the plant spring from ground touched by the serpent as it was driven out of the Garden of Eden. The Egyptians were among the first to use it as a "vermifuge" to expel parasitic intestinal worms, giving the herb its common name "wormwood". It was also used to relieve pain "of demonic origin" in the anus.

Artemisia (named for the goddess Artemis) was recommended by Hippocrates for menstrual pain and jaundice, by Dioscorides as a general health drink and by Galen for stomach relaxation, but it was also burned with sandalwood to contact the dead. According to Pliny the Elder, champion chariot racers were given a drink made with *Artemisia absinthium* to remind them that glory may also bring bitterness.

Wormwood was a powerful weapon in the herbalist's armoury. Apart from the classic use where, as John Gerard so memorably noted, it "voideth the wormes of the guts", it could combat poisons and, most usefully of all, deter sea dragons. Culpeper recommended wormwood's antiseptic properties for stings by bees, wasps, scorpions and snakes. He also suggested that, in wardrobes, it would "make a moth scorn to meddle with the cloaths as much as a lion scorns to meddle with a mouse or an eagle with a fly".

Combined with mugwort (*Artemisia vulgaris*), it was rumoured to conjure spirits. Of course, those spirits might be of the alcoholic variety. The mysterious W.M., allegedly former cook to Queen Henrietta Maria, includes in *The Queen's Closet Opened* (1655) a recipe for candied wormwood, but it was generally considered a dangerous herb. After mixing it with water, wine and brandy, it was only a matter of time before wormwood became the star of its own spirit. Highly addictive, and with mind-altering effects, unregulated absinthe became as much of a scourge in nineteenth-century France as gin was in England. A nationwide ban, imposed in 1914, was only lifted in the early twenty-first century. A regulated version now enjoys EU geographically protected status.

Opposite Wormwood (*Artemisia absinthium*) from Köhler's *Medizinal Pflanzen*, 1887.

Artemisia Absinthium L.

SOLANACEÆ

Hyoscyamus niger (L.)
Henbane
NATURAL ORDER Solanaceae
DATE July 14th 1895
HABITAT Waste place nr. Folkestone.

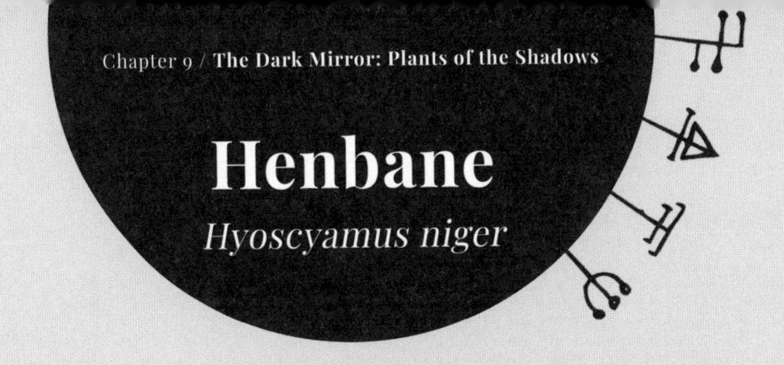

Henbane

Hyoscyamus niger

In 1910, the notorious murderer Dr Crippen was convicted of using the powerful narcotic hyoscine to poison his wife.

The drug was found in the body, the doctor had bought it, and had no convincing argument for having so much of the toxin in his collection of homeopathic medicines.

Hyoscine is derived from one of our most deadly plants: henbane. It causes delirium, loss of speech and paralysis. Known for growing around middens, "the devil's eye" was also known as "stinking roger", "henpenny" and "hog's-bean". It even looks suspect: woolly leaves, flowers with purple veins that emulate decay and a thick, white taproot that emits a noxious, foul-smelling smoke when burned. But, as herbalists always know, harm can be harnessed. Dioscorides believed the smoke had medicinal value as an analgesic and it was used by the Romans to alleviate pain in childbirth.

Medieval churchmen were more circumspect, suspecting that witches burned henbane to conjure spirits and gain powers of clairvoyance, though other people thought it could counteract witch-harm.

It was most useful as a painkiller, albeit a dangerous one. Country folk were known to smoke henbane like tobacco as a cure for toothache, calculating that the convulsions they risked couldn't be any worse than the pain. This led to a notorious confidence trick, of which John Gerard is scathing. Quack dentists would direct henbane smoke into the mouths of patients, telling them it would kill the worms that were eating the affected tooth. The pain was temporarily numbed and the doctor's assistant handed over a cup with which to rinse. The patient would be delighted when they spat, not knowing the "dead worms" were actually chopped-up lute strings.

Nicholas Culpeper was nervous around henbane, though he did admit that, applied externally, it could be useful. A warm fomentation, for example, would ease swellings of the testicles or women's breasts. The leaves, boiled in wine, could soothe inflammation of the eyes, gout, sciatica and general joint pain while, mixed with vinegar and applied to the temples, henbane could relieve a headache. The same mixture would cure insomnia if used to bathe the feet.

It was never to be taken inwardly. Culpeper listed some possible antidotes for anyone unlucky enough to have been poisoned with it – goat's milk, honey, water, mustard seed – but was not optimistic about the outcome.

Opposite Herbarium sheet of henbane (*Hyoscyamus niger*) collected at Kew, 1895.

Deadly nightshade

Atropa bella-donna

Deadly nightshade is a pretty plant – far too pretty. Its deep purple bell-shaped flowers are stunning and the juicy green-red-black berries look sweet and tasty.

Don't be fooled. Everything you need to know about *Atropa bella-donna* is in the names, both common and Latin.

Atropos is the eldest of the three Greek Moirai, or Fates. After her first sister, Clotho, spins the thread of human life and her second sister, Lachesis, measures its length, Atropos "the inflexible" cuts it off. Her Roman equivalent is Morta, the goddess of death.

Bellona, who has also been associated with the plant, was a Roman goddess of war, but the second part of deadly nightshade's Latin name comes from the habit practised by Venetian women of putting drops of the herb into their eyes. Their pupils dilated, supposedly making them look more attractive: *bella-donna* ("beautiful woman"). In reality, they probably looked more spaced-out than doe-eyed, thanks to atropine, a sedative found in the plant.

Atropine causes severe sweating, vomiting, difficulties in breathing, hallucinations and, possibly, coma followed by death. Deadly nightshade may belong to the same Solanaceae family as tomatoes, potatoes, aubergines, chillies and peppers, but every part of this plant is toxic. Indeed, that is precisely why people were suspicious of foods like tomatoes when they were first introduced to Europe.

There is relatively little folklore associated with deadly nightshade. It was important to attach no romance whatsoever to it – this one was a killer. Children, most at risk of seeing the shiny berries as worth putting in their mouths, were told they would meet the devil – or even death itself – if they picked them. This was true – three berries was considered enough to kill a child.

Deadly nightshade is said to be the "insane root" mentioned in Shakespeare's *Macbeth* and it was also said to be one of the strongest magical ingredients in witches' flying ointment. It has been used to make poison-tipped arrows and, allegedly, to kill Roman emperors, from Augustus to Claudius. Despite all this, the plant is used in tiny amounts in some modern medications.

Deadly nightshade should not be confused with woody nightshade (*Solanum dulcamara*) whose purple star-shaped flowers and tomato-like berries hang in clusters. Woody nightshade is not as poisonous as its cousin, though it will cause a nasty stomach ache; anyone eating it will probably need medical attention.

Opposite Deadly nightshade (*Atropa bella-donna*) from *English Botany*, James Sowerby, 1791–1814.

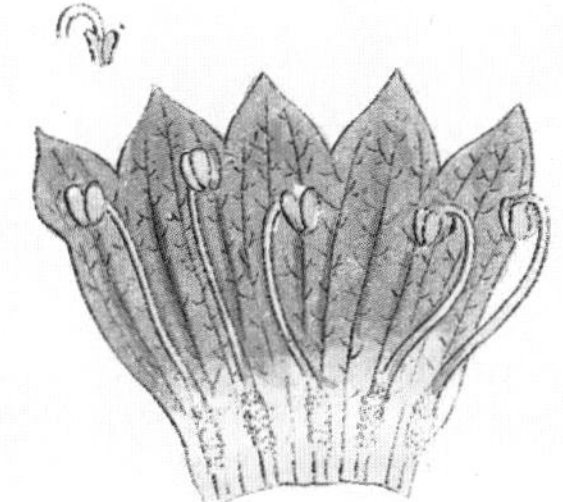

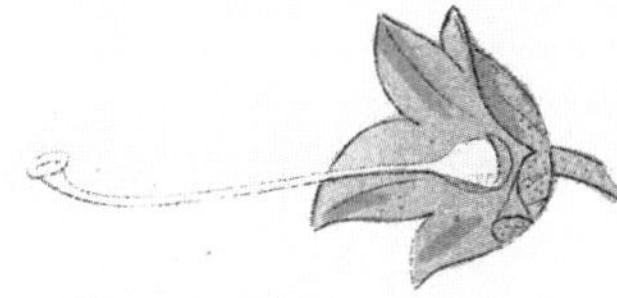

B. 592.

Atropa Belladonna. Deadly Nightshade.

Wallich 1828
East Ind. Co.
Aconitum ferox, Wall.

Monkshood

Aconitum napellus

There are at least 100 species of monkshood, also known as aconite. None of them are friendly.

A member of the buttercup (Ranunculaceae) family, every part of this hairless perennial, with its segmented leaves and blueish-violet flowers, is poisonous. Monkshood poisoning causes stomach pain and dizziness, and can be fatal. It also affects the heart, but happily, it tastes so awful that accidental poisoning is unusual. Humans have been in awe of this most dangerous of plants since antiquity.

The herb was said to grow around Hades, the god of the Greek Underworld, springing up from the spittle of his three-headed dog, Cerberus. The jealous goddess Athena sprinkled the maiden Arachne with aconite juice, turning her into a spider. It was even said that on the Greek island of Chios, senile men were given aconite as a form of euthanasia.

In Hindu mythology, the plant is sacred to Shiva, who saved the world by drinking poison. The experience turned the god blue and a few drops that missed his mouth fell to the ground as aconite. Norse folklore claims the plant for Thor; other northern European traditions associate it with Hecate, which is why it was found at crossroads and gateways, some of her remits as goddess.

Taking monkshood is said to lend a tickling sensation to the skin, as though the drinker is growing fur. This may account for the hardcore Viking warriors that called themselves "berserkers" (from *ber*, meaning "bear", and *serkr*, meaning "shirt") allegedly consuming "wolfsbane" to transform themselves into werewolves. The Latin name for another aconite, *Aconitum lycoctonum*, means "wolf-killer". One fiftieth of a grain, it was said, would kill a sparrow; Anglo-Saxon hunters tipped their arrows with monkshood before chasing wolves. In certain parts of Alaska, it was used to poison whale harpoons.

To the medieval church, any plant as dangerous as this had to be involved in witchcraft. Rubbed into the skin, it produced numbness; ingested, it made the heart race; and combined with deadly nightshade (*Atropa bella-donna*), the "devil's helmet" caused delirium, enough to give the mental sensation of floating, making monkshood a classic ingredient of flying ointment.

Culpeper was, understandably, nervous about prescribing the plant, though considered one yellow-flowered species he called the "wholesome wolf's bane" (*Aconitum anthora*) useful in a lotion for venomous bites.

Opposite Indian aconite (*Aconitum ferox*), 1828, from the Indian Plants section of Kew's Wallich Collection.

Blackthorn

Prunus spinosa

Blackthorn is one of the few hedgerow plants many of us still forage – or at least those of us counting ourselves gin lovers.

The blackthorn's powdery, black "sloes", sharp to the taste but a perfect foil for spirits, are a classic gin flavouring.

With a name like "blackthorn", this woody, thorny, "wild cherry" should have a terrible reputation, but although it *is* sometimes associated with witchcraft, it is also a lucky plant, sometimes even a horticultural hero of fairy tales. Irish folklore has heroes throwing a blackthorn stone behind them to create a hedge, delaying evil pursuers. Some versions of *Sleeping Beauty* have a similar blackthorn hedge though it can be less helpful: *Rapunzel's prince* is blinded by sharp spines sometimes attributed to blackthorn.

Like so many traditional hedging plants, it has both good and bad connotations. The "dark crone of the woods" was the keeper of secrets, perhaps based on Cailleach, the Celtic queen of winter. Yet the simple flowers, reminiscent of cherry blossoms (and white, so unlucky indoors), were also supposed to bloom on old Christmas Day, in January. In some regions, a "blackthorn winter" – a snap of unseasonably cold weather after spring has begun – was thought a good time to sow barley.

Sloe thorns were thought to be used by witches to prick wax images. In Scotland in 1670, Major Thomas Weir was burned as a witch, along with his blackthorn staff. His ghost still carries the stick. Care needed to be taken with accusations, however, as blackthorn wands were also badges of high office. The most famous is carried by the Gentleman or Lady Usher of the Black Rod, a Sergeant at Arms to Britain's House of Lords. The role was created in 1348 by Edward III to keep the peace between unruly Knights of the Garter. The first incumbent carried a simple rod of blackthorn; today's is fancier. In Ireland, fearsome *shillelaghs* (fighting clubs) were made of blackthorn root. Less lethally, the wood also sometimes made divining rods.

Powdered blackthorn bark eased fevers, but most uses relied on juice from the velvety black fruit. Generally picked after the first frosts to break down the skin, sloes made a rich purple-pink dye. The ancient Greek physician Andromachus prescribed the juice for dysentery and Culpeper agreed it made a good remedy for the fluxes of the bowels. Highly astringent, it was also used as a mouthwash for loose teeth and acted as a diuretic purge. Too much of the juice can be toxic, but since raw sloes are very bitter, accidents are rare.

Opposite Blackthorn or sloe (*Prunus spinosa*), illustrated by O.W. Thomé in *Flora von Deutschland Österreich und der Schwei*, 1885.

XII,1.
105. Rosaceae. 1. Pruneae.
1
2
3
4
A
5
6
9
10
B
7
8
394. Prunus spinosa L.
Schlehdorn.

AGARICINÉES.

AMANITA MUSCARIA. Fr.

½ Gr. nlle.

Fly agaric

Amanita muscaria

Fungi are strange. Neither animal nor vegetable, they are the fruiting bodies of much larger organisms spreading through soil or organic substances, from dead trees to animal skin.

Because of the mysterious way they appear and disappear, often forming "fairy rings", many fungi were thought to grow where lightning had struck and were often considered sacred.

Some poisonous fungi masquerade as edible. The death cap (*Amanita phalloides*) is one of the world's great toxins, but appears perfectly normal when young.

Other fungi don't even pretend to be edible. Fly agaric was sometimes said to have been the food of the gods. It certainly wasn't for humans. The quintessential "fairy" toadstool, it fits many descriptions of divine foods in ancient texts. The Hindu "*soma*", for example, an unidentified plant sung about in the hymns of the *Rig Veda*, was pressed for its juice, filtered through sheep's wool and offered to the gods. Some have argued fly agaric is the main ingredient in "ambrosia", nectar of the Greek gods.

The Koryak people of Siberia tell of the folk hero Big Raven, whose brother Whale became stuck in mud. The sky god Vahiyinin (Existence) told him to eat "wapaq" (earth spirits), which would be wearing little red hats with white spots. Big Raven found the wapaq in the form of fly agaric, which he ate. Now possessing the strength to fly, he carried Whale back to the sea in a giant bag. Big Raven was so impressed with the fungus's usefulness that he caused it to stay, so his children could discover cures for ailments or the meaning of dreams by eating it too. Fly agaric is a strong hallucinogen so, as a safety measure, a human shaman ingested the mushrooms first, then everyone else drank his urine.

It's been suggested that fly agaric explains the uncontrolled behaviour of Viking "berserkers" and that followers of the Roman cult of Mithras used the fungus for its hallucinatory properties.

They would have needed to have been brave. The fungus is found in woodland, including on beech, birch and conifers, fruits in late summer and autumn until the first frosts. Even dried, it is powerful stuff. Containing the psychoactive compound ibotenic acid and traces of the sedative and hypnotic muscimol, ingestion can be very unpleasant, and any vivid dreams and sensations of weightlessness are cancelled out by less pleasurable symptoms like incoherent speech, vomiting, diarrhoea, rapid breath, slowed pulse, dizziness, drowsiness, headaches, seizures, delirium and, possibly, coma or even death.

Opposite Fly agaric (*Amanita muscaria*), a hallucinogenic and toxic fungus.

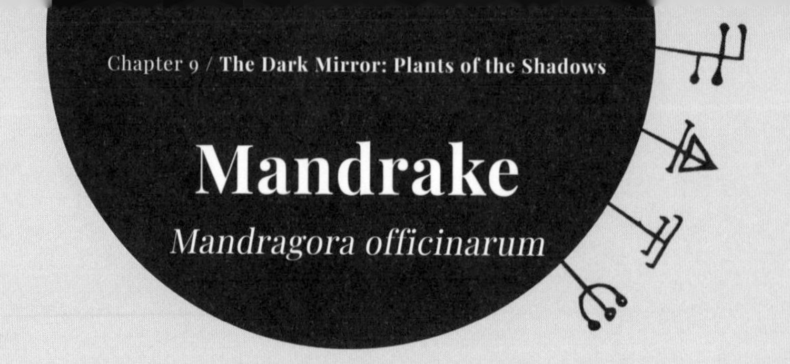

Mandrake

Mandragora officinarum

Of all the plants in a medieval herbal, the mandrake is easily the most iconic, and not just to *Harry Potter* fans.

It's not so much that this rosette-leaved, purple-flowered, orange-fruited plant has a taproot in the shape of a "human", more that medieval artists loved illustrating it. There are dozens of visual interpretations of "Satan's apple" (also "devil's turnip"). It was said the scream of a mandrake when pulled would drive the harvester insane, so a dog should be tied to it, then scared into running away, pulling up the root. It was, of course, a story put about by professional root-cutters worried about losing business.

Containing hyoscine and hyoscyamine, the ancient Greeks used it as an anaesthetic. The mandrake was said to have sprung from the dripping "juices" of a hanged man, whose likeness forever appears in the root. Under the doctrine of signatures, mandrake root looked a bit like an incredibly ugly baby, so sleeping with a piece under one's pillow aided conception. The roots also made general good-luck talismans which, if the right person asked, would whisper the location of buried treasure. Once over his disappointment, Culpeper found the leaves "cooling" and the root a powerful emetic that "few constitutions can bear".

Mandrakes weren't so easy to come by, however. "Venus nights" in the fens region of England saw competitors vie to see who could dig up the root that looked most like a human female, but others turned to more desperate measures.

Shysters did a roaring trade at county fairs, selling "mandrakes" to childless but hopeful women. These were usually fakes, carved out of misshapen turnips. Sometimes they were even stuffed with grass seed so they would magically sprout "hair". Other charlatans made crude clay figures around a plant's roots and reburied them so they could miraculously "discover" a mandrake. Anyone fooled by them might hope the fraudsters would fall into the hole left by the root, which allegedly went straight to Hades.

The English mandrake (*Bryonia cretica* subsp. *dioica*) is a hedgerow climber, a completely different – and much more deadly – plant. Culpeper warns, "it is not rashly to be taken", but he felt it useful for falling sickness (epilepsy), palsies, convulsions and dropsy, if properly dispensed. It was also an abortifacient. Neither plant is recommended for use today.

Opposite Mandrake (*Mandragora officinarum*) from *Hortus Sanitatis*, Jacob Meydenbac, 1485.

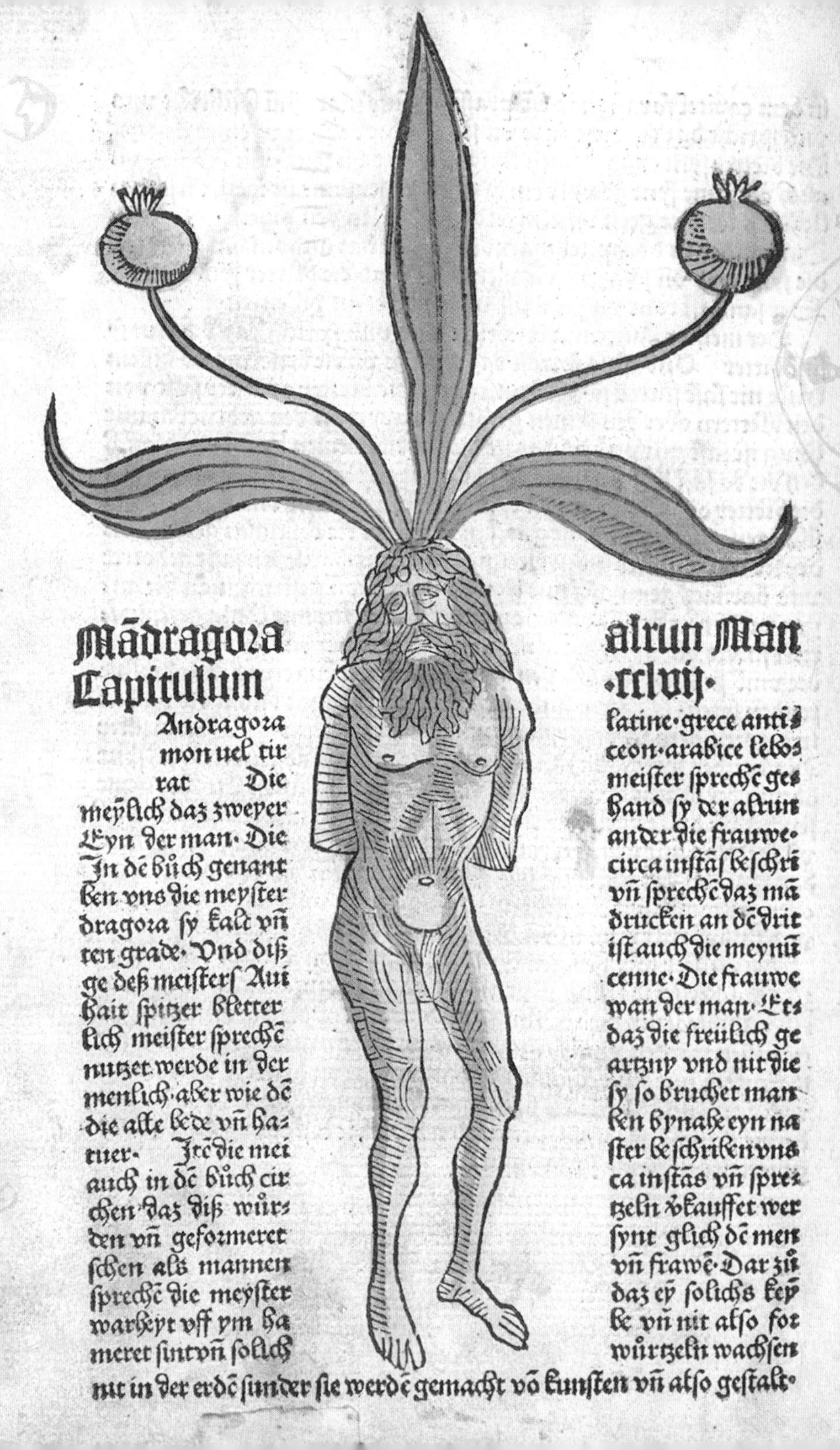

Mādragora alrun Man
Capitulum ·cclvij·

Andragora
mon uel tir
rat Die
meÿslich daz zweyer
Eyn der man· Die
In dē bůch genant
ben vns die meyster
dragora sy kalt vn̄
ten grade· Vnd diß
ge deß meisters Aui
hait spitzer bletter
lich meister sprechē
nutzet werde in der
menlich· aber wie dē
die alte bede vn̄ ha
tuer· Jtē die mei
auch in dē bůch cir
chen daz diß wůr
den vn̄ geformeret
schen als mannen
sprechē die meyster
warheyt vff ym ha
meret sint vn̄ solich

latine· grece anti
ceon· arabice lebos
meister sprechē ge
hand sy der alrun
ander die frauwe·
circa instās beschri
vn̄ sprechē daz mā
drucken an dē drit
ist auch die meynū
cenne· Die frauwe
wan der man· Ets
daz die freülich ge
artzny vnd nit die
sy so bruchet man
ken bynahe eyn na
ster beschriben vns
ca instās vn̄ spre
tzeln v̄kauffet wer
synt glich dē men
vn̄ frawē· Dar zů
daz eӱ solichs key
be vn̄ nit also for
würtzeln wachsen

nit in der erdē sunder sie werdē gemacht vō kunsten vn̄ also gestalt·

Chapter 10:

Plants of Hope

Folklore has many flowers of ill omen, but just as many plants signal good luck and hope. From herbs that bring lovers together to those lending solace to the grieving, these are flowers of joy.

Spring in itself brings hope – of new life and warmer days ahead. At this time we look to the natural world for plants that bring joy and herald the coming fruitful months, where winter is a distant memory.

Traditionally, plants that flowered in the early months were good omens, including the delicate blossoms of fruit trees bursting from the bare wood. Less delicate than it appears, the snowdrop (*Galanthus*), when it bursts through the frozen earth, is known in France as "*perce-neige*" ("snow-piercer"). A symbol of purity and strength, it brings hope today, too, in the form of galantamine, a prescription drug based on alkaloids found in members of the snowdrop family and used to treat patients suffering from Alzheimer's disease.

To the Māori people, ferns represented new life and new beginnings – it is still the emblem of New Zealand's All Blacks rugby team. In Japan, ferns bring hope for future generations in a family.

Clover (*Trifolium*) is associated with the Irish St Patrick, who used the three-leaved shamrock (or lesser trefoil, *Trifolium dubium*) to teach his followers about the Holy Trinity. When a clover sprouted four leaves, however, it became a magical plant. Tradition, from Scandinavia to Eastern Europe, suggested it grew where a mare dropped her first foal or where a foal had first sneezed. There was no point purposefully looking for such a plant; it could not be found by searching, but accidental finders could see the fairies, repel witches and break enchantments. In the English county of Cambridgeshire, girls could indulge in the risky practice of placing a four-leafed clover in their shoes; they would marry the first man they met. Today, it's possible to buy "clovers" whose leaves are quatrefoil, but they are usually hybrids of the wood sorrel (*Oxalis*).

Not all plants of hope are those of the spring and summer. In China and Japan, the chrysanthemum is highly revered. One legend tells how a spirit informed a bride that her marriage would last as long as the number of petals on the flower she wore at her wedding. She chose the chrysanthemum, then divided each petal, acquiring 68 years of married bliss.

The most famous symbol of hope is the olive branch, brought to Noah by a dove, as a peace offering from God after the great flood of Genesis. According to pagan tradition, the first olive tree, *Olea europaea*, was given to the Greeks by the goddess Athena. In 480 BCE, after the Battle of Thermopylae, the acropolis was burned down by Xerxes's army. All hope was lost. But the next day, there were fresh buds on the charred branches of the sacred olive tree. Its seeds were distributed far and wide, bringing promise of better times to the Greek nation. May all plants bring such happy thoughts.

Opposite The olive (*Olea europaea*) is perhaps one of the most universally-acknowledged symbols of peace and reconciliation.

OLEA, OLIVIER. N°. 17.
Page 88 **

a. Trifolium montanum flore albo. b. Trifolium montanum spica longissima. c. Trifolium pratense album. d. Trifolium pratense rubrum. e. Trifolium pratense folliculatum. f. Trifolium vesicarium purpureum. g. Trifolium siliquosum.

x

Opposite White clover (*Trifolium repens*) from *Phytanthoza iconographia*, Johann Wilhelm Weinmann, 1737. It was said to repel fairies and witches.

Above Chrysanthemum (*Chrysanthemum indicum*) from *Gartenflora*, 1897. It is associated with happy marriages in Chinese legends.

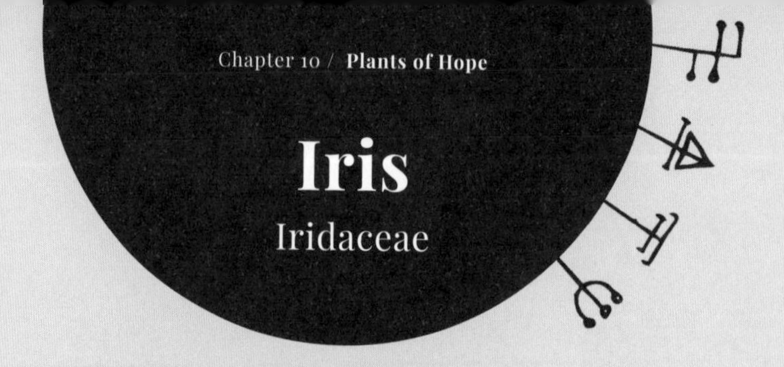

Iris

Iridaceae

The main job of the Greek goddess Iris was attending Hera, but no one remembers her for that.

She is far better known as the personification of the rainbow and, as a messenger goddess, usually represented in art with a fine set of wings. The other gods found her a little intimidating, as she carried the water that held them to their most solemn oaths. But mortal women revered her as their guide in the Underworld. For this reason, the flowers bearing her name were planted around ancient Greek tombs as a symbol of hope.

The Iridaceae family is large, including crocuses and freesias as well as a wide variety of irises, from the small, wild varieties often found in bogs or on the margins of water to the larger, blousier, bearded *Iris* x *germanica.* A popular plant since antiquity, the iris is pictured in Egyptian tombs and used by the ancient Greeks and Romans.

Whether the marginal "yellow flag" (*Iris pseudacorus*); the splendidly named "stinking gladwyn' (from the old English word *gladwyn*, or "sword") *Iris foetidissima* (also known as the "roast-beef plant"); or one of the larger species, most irises share long, blade-like leaves, flowers on long, leafless stalks and fleshy, creeping rhizomes.

Although used in the past as a cure for snake bites, coughs and scrofula, and chewed as a cure for bad breath, iris always seems to have been a bit of a medical afterthought. The iris's true calling was as orris root. Discovered by the Greeks and Romans, orris is the dried rhizomes of (usually) *Iris pallida* and *Iris germanica*, and is still heavily used by the perfume industry. The ancients used it in face and body powders; by medieval times, it was being sprinkled into linen. Today, it is used in high-end fragrances, dry shampoo and even as a botanical in trendy gins. It has always been punishingly expensive, as the rhizomes are odourless when first harvested. It takes three to four years for the dried bulbs to reach their full, sweet, floral pungency (a smell similar to that of Parma violets). Sometimes orris is further distilled into orris oil – which, as well as smelling wonderful, also "fixes" other, less stable aromas.

The iris is also said to have been the inspiration for the famous heraldic fleur-de-lis. Legend tells that Louis VII, King of the Franks, dreamed of irises before setting out on a crusade in 1137. The resulting "fleur de Louis" became "fleau de luce" and, finally, "fleur de lis". Some claim it represents prosperity, royalty and valour, but more often, the three petals are thought to be symbols of faith, charity and hope.

Opposite *Iris sofarana* from *Calendarium*, Sebastian Schedel and Basilius Besler, 1610.

Endnotes

1 Spells and charms from the *c.* seventeenth-century *Book of Magical Charms*, held at and transcribed by the Newberry Library, Chicago.
2 Ibid.
3 Ibid.
4 Ibid.
5 Ibid.
6 Thompson, H. "How Witches' Brews Helped Bring Modern Drugs to Market". *Smithsonian Magazine*, 31 October, 2014. https://www.smithsonianmag.com/science-nature/how-witches-brews-helped-bring-modern-drugs-market-180953202/. Accessed 4 June, 2020.

Bibliography

To list all the books and articles referenced for this work would be impossible but here is a list of the volumes I found most consistently useful. A lot of these are, alas, out of print but can be found reasonably easily second-hand.

Vickery's Folk Flora (Orion, 2019) is a magnificent life's work by botanist/folklorist Roy Vickery. Vickery is also the brains behind **Plant-lore.com** and, the Folklore Society's 1985 gem, ***Unlucky Plants***.

Margaret Baker is the author of several important folklore books, including ***The Gardener's Folklore*** (David & Charles, 1977). Her ***Discovering the Folklore of Plants*** (Shire Publications, 1969) is a tiny book, packed full of excellent information.

Anything with Chris Howkins's name on it is worth reading. His self-published series of slim volumes often discuss individual trees, such as ***Rowan, the Tree of Protection*** (1996), ***Elder, the Mother Tree of Folklore*** (1996), and ***Holly, a Tree for All Seasons*** (2001) but I also found ***A Dairymaids' Flora*** (1994) and ***Valuable Garden Weeds*** (1991) invaluable.

Ruth Binney's work is meticulous and fascinating. I will read anything she writes but I am particularly grateful for ***Plant Lore and Legend*** (Rydon, 2016).

Nial Edworthy's ***The Curious Gardener's Almanac – centuries of practical garden wisdom*** (Eden Project, 2006) was, indeed, a wise read.

A Garden of Herbs by Eleanour Sinclair Rohde was originally published in 1936 but reprinted by Dover in 1969. It is packed full of fabulous folklore, history and receipts.

Plants of Mystery and Magic, A Photographic Guide by Michael Jordan (Blandford, 1997) provided a good reference for some plants not mentioned elsewhere, as did the sadly anonymous ***Plant Folklore Pocket Reference Digest*** (Geddes and Grosset, 1999).

Bill Laws's ***Spade, Skirret and Parsnip*** (Sutton, 2004) was an excellent source of information about odd vegetables, while Brigit Boland's ***Old Wives' Lore for Gardeners*** and ***Gardener's Magic and Other Old Wives' Lore*** (Bodley Head, 1977) are good for the really strange stuff. ***Ireland's Wild Plants: Myths, Legends and Folklore*** by Niall Mac Coitir (The Collins Press, 2015) is a fascinating read.

Nicholas Culpeper's ***Herbal*** has seen many editions. A truly important "basic" herbal, I could not have done without it. John Gerard may have been a bit of a rogue but the ***Herball*** that bears his name is still very useful.

Medicinal Plants of the World by Ben-Erik van Wyk and Michael Wink (Timber, 2004) was also invaluable as was Richard Mabey's glorious ***Flora Britannica*** (Sinclair-Stevenson, 1996).

I consulted many, many books on general folklore, from regional tales to volumes around the world. Two classic books on superstitions, E & M.A. Radford's ***Encyclopedia of Superstitions***, (Book Club, 1961) and the inestimable Steve Roud's ***Penguin Guide to the Superstitions of Britain and Ireland*** (Penguin, 2003) are excellent reads as is Roud's ***The English Year*** (Penguin, 2007).

Index

Picture credits

The vast majority of images in this publication are reproduced from the archive collection of the Royal Botanic Gardens, Kew, with the exception of the following:

Alamy: Charles Walker Collection 42-43; /Classic Image 62; /Heritage Image Partnership 87, 153; /Les Archives Digitales 4-5, 6, 20, 34, 58, 76, 96, 127, 134, 146, 149, 168, 179, 194; /Pictorial Press 103, 178; /Science History Images 64, 99; /World History Archive 140

Bridgeman Images: Bibliothèque Nationale, Paris, France 83; /British Library Board. All Rights Reserved 139

Getty Images: Culture Club 172; /Prisma/UIG 39

Public Domain: 14, 38, 104, 174

Wellcome Library: 37, 48, 49

Every effort has been made to acknowledge correctly and contact the source and/or copyright holder of each picture and Welbeck Publishing apologises for any unintentional errors or omissions, which will be corrected in future editions of this book.